ANTARCTICA'S
HIDDEN HISTORY
CORPORATE FOUNDATIONS OF SECRET SPACE PROGRAMS

ALSO BY DR. MICHAEL SALLA

The U.S. Navy's Secret Space Program & Nordic Extraterrestrial Alliance
— Book Two of the Secret Space Programs Series —

Insiders Reveal Secret Space Programs & Extraterrestrial Alliances
— Book One of the Secret Space Programs Series —

Kennedy's Last Stand:
Eisenhower, UFOs, MJ-12 & JFK's Assassination

Galactic Diplomacy
Getting to Yes with ET

Exposing U.S. Government Policies on Extraterrestrial Life

Exopolitics:
Political Implications of Extraterrestrial Life

ANTARCTICA'S
HIDDEN HISTORY
CORPORATE FOUNDATIONS OF SECRET SPACE PROGRAMS

Michael E. Salla, Ph.D.

Hawaii, USA

ANTARCTICA'S HIDDEN HISTORY
CORPORATE FOUNDATIONS OF SECRET SPACE PROGRAMS
Copyright © 2018 by Michael E. Salla, Ph.D.

Exopolitics Consultants
RR2 Box 4876
Pahoa, HI 96778 USA

Printed in the United States of America

Managing Editor: Angelika Whitecliff
Copy Editor: A. Hughes
Cover Design: Rene McCann

ISBN 978-0-9986038-2-7

Library of Congress Control Number: 2018902976

Author's website: www.exopolitics.org

CONTENTS

TABLE OF FIGURES

PREFACE

A spotlight was cast on many previously unknown historical events by the revelations of William Tompkins, a brilliant young naval recruit who became a career aerospace engineer. Among his notable list of disclosures was that the US Navy had established a top secret espionage program from 1942 to 1946, with approximately 30 spies embedded within Nazi Germany's top aerospace facilities and companies. He also reported that Nazi Germany had secretly developed up to 30 prototype spacecraft using antigravity and other exotic propulsion systems. Having sat in on the debriefings of the spies in his capacity as "Disseminator of Naval Research and Information", Tompkins attests that many of these prototypes were extraterrestrial in origin, and the most technologically advanced were secretly being developed within hidden bases in Antarctica. Furthermore, he claims that the Nazis were helped by two different groups of extraterrestrials in achieving numerous stunning technological breakthroughs in their nascent aerospace industry, which was relocated early on to Antarctica.

A yet untold story emerges with this book's investigation of the role of German companies and their U.S. partners in the funding and development of antigravity spacecraft in Nazi Germany. This involves some of the biggest companies in Germany, as well as major U.S. corporations who provided critical financial services and technological assistance. Some of the U.S. corporate officials directly involved in the funding and technological development in Germany rose to very senior positions within the U.S. political system after World War II. They include Allan Dulles, who became CIA Director (1953 – 1961); John Foster Dulles, who ascended to the role of Secretary of State (1953-1959); and Prescott Bush, who came to be a U.S. Senator (1952 to 1963), and is widely remembered as the father and grandfather of two future U.S. Presidents who would enthusiastically continue the legacy of Prescott's policies. Finally, we have the Rockefeller brothers who provided important corporate support to Nazi Germany, and became a prominent part of the Eisenhower administration through Nelson who filled various senior positions.

These men, along with other U.S. corporate/government officials, are directly implicated in aiding the success of specific German companies both prior to and after World War II, and gaining them the financial and international corporate support necessary for the building of a secret space program in Antarctica. Perhaps most alarming is that these prominent officials made it possible for a post-World War II agreement to be reached between the Eisenhower Administration and a German breakaway group in Antarctica which came to establish the vision of a Fourth Reich; whose power and influence extends to present day.

All the resources and manpower of the U.S. military-industrial complex were hence made available for a significant expansion of the German space program, whereby it could become an interplanetary colonial power capable of deploying significant military resources outside of our solar system in support of their extraterrestrial allies. Furthermore, it made possible the development of a transnational corporate space program, whose power and influence exceeded the military space programs established by the Navy and Air Force respectively within the U.S.

In the following chapters, I will detail and expose how German secret societies and companies collaborated in building their first prototype spacecrafts, and how these led to the fully operational and weaponized spacecraft in Antarctica. Key elements of this story include the Nazi *capital flight* that funded these secret technological developments even after the collapse of Hitler's Third Reich, and how the spacecraft from Antarctica first defeated a U.S. Naval expedition sent out to find and destroy the German Antarctica bases in 1946/1947, and later, in a stunning display of power flew over Washington, DC in July 1952 to intimidate the Truman administration into beginning secret negotiations. A pivotal moment in history resulted with the agreement between the subsequent Eisenhower administration and the Germans in Antarctica, thus heralding the emergence of the Fourth Reich as a global superpower that would remain unknown to the majority of the world's population.

I next follow the telling trail of the extensive use of slave labor that was first adopted by the political leaders and corporations building the space program in Germany during WWII, which ostensibly continued in Antarctica. This egregious practice was then adopted by U.S. companies that partnered with the Germans in Antarctica to develop more advanced spacecraft, which in turn would be used for deep space operations and to even establish colonies within the solar system.

The intriguing connection between the 1950's Space Brothers phenomenon and the German Space Programs out of Antarctic will be examined in chapter seven. Importantly, President John F. Kennedy was aware of the U.S. agreement with the German Antarctica space colony. His attempt to assert direct Presidential authority over these secret agreements and the technologies they involved became a direct factor leading to his assassination, which will be divulged in detail in chapter eight.

Of the many German companies that helped build Nazi Germany's secret space program(s), Siemens stands out as the most significant, as will be discussed in chapter nine. It was by far the most successful German company across a wide spectrum of research and development projects related to multiple components of antigravity spacecraft, and prototype craft. These spacecraft prototypes were eventually redeployed to Antarctica, along with technologies from Siemens' subsidiaries and other German companies that had successfully built key components for space faring vehicles. Based on its overall success in achieving advanced breakthroughs, strategic business partnering and acquiring ample funding, Siemens was able to play a leading role in coordinating German corporate construction of multiple spacecraft in Antarctica.

Siemens was also implicated in managing the slave labor component of the German Antarctica program, thereby continuing the corporate policy that it had established in Nazi Germany under the encouragement of the Nazi regime. This continued decades after World War II ended, and led to Siemens secretly manufacturing billions of RFID tracking chips in the 1980's to allegedly track the slave laborers used in their German-U.S. Corporate Space Program that by then had established large operations in Antarctica, and also on the Moon, Mars and elsewhere in the solar system. In addition, Siemens and other corporations working in the German-U.S. corporate space program (aka, Interplanetary Corporate Conglomerate), are linked to the abduction of humans used in a galactic slave trade with extraterrestrial civilizations. Remote and with perilous conditions, Antarctica would become a key outpost for this burgeoning slave trade.

Chapter 10 identifies how the original German Space Program and the transnational corporate space program it spawned are kept safely secret and secure under the depths of Antarctic ice. Few scientists and visitors to Antarctica ever witness anything that makes them suspect what is really happening under the massive frozen expanse. Those who do witness anomalous events, such as Brian, a retired Navy aircraft engineer, are debriefed to never reveal what they saw. Those brave enough to ignore such warnings and step forward as whistleblowers, which Brian did, are then later threatened by strangers to keep silent.

In addition to the Antarctica revelations of William Tompkins, we also have the testimony of Corey Goode, who says he served in a "20 and back" tour of duty in the US Navy's secret space program, Solar Warden. His claims of serving in Solar Warden, and having extensive contact with multiple groups of extraterrestrials, were extensively examined in the first volume of this Secret Space Programs Series.[1] In order to assist the reader in understanding the different space programs that have allegedly been developed according to Goode, I've included a reference diagram developed by aerospace engineer and former NASA employee, Thomas L. Keller, summarizing the testimony of Goode. In order to better follow the information in this book when different space programs are discussed, I recommend readers refer to this diagram.

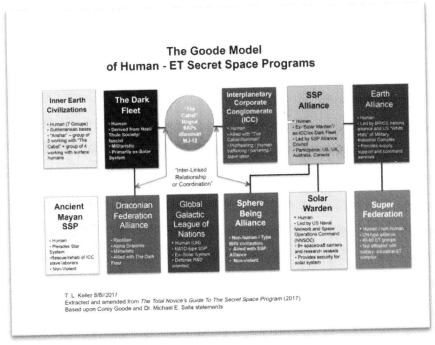

Figure 1. Illustration of different Space Programs according to Corey Goode.

Goode says he was taken twice to Antarctica between January 2016 and early 2017 – by an Inner Earth group he identifies as the Anshar. He has provided extensive testimony about what he saw in terms of large industrial facilities hidden deep under the ice shelves, where powerful fleets of spacecraft are based and assembled. If Goode's testimony is reliable, as multiple sources of evidence suggest, then we have the first eyewitness account of the transnational corporate space program currently operating in Antarctica.

In addition, Goode says that he witnessed the remnants of an ancient civilization being secretly excavated under the Antarctic ice. His description of the "Pre-Adamites" discovered there, along with their highly developed technological society, is stunning in its implications for understanding our ancient human history. Chapter 12 also examines scientific data that shows Antarctica's volcanoes waking up, and the potential this has to melt the ice shelves, revealing what lays hidden beneath. The connection of the Pre-Adamites to different historical epochs in our solar system, and the role of Antarctica as a refuge for escaping Martians and the inhabitants of a former Super Earth from the asteroid belt is laid out and scrutinized in chapter 13.

The connection between Antarctica and the Fallen Angels described in the Book of Enoch ensues and is analyzed in chapter 14. It will be shown how this apocryphal book points to Antarctica as the location of their imprisonment. Most incredible is Goode's claim that some of these Pre-Adamites are still alive in stasis chambers and that their bloodline descendants want to not only awaken them, but also restore them to their previous positions of authority. Indeed, many alarming questions are raised if only a portion of Goode's claims are accurate.

Next, the penultimate chapter examines secret military research and development that has occurred in Antarctica, which has gone forward despite the 1961 Antarctic Treaty proscribing such activities. Consequently, while Antarctica is ostensibly a region governed by the Antarctic Treaty establishing a demilitarized zone for scientific exploration to benefit all humanity, it is in fact a heavily militarized territory conducting many illegal research and development programs involving captive humans.

This book exposes how human slavery continues to flourish in Antarctica and on off-planet colonies run by the Fourth Reich and their transnational corporate partners. If left undisturbed, this nexus of domination and slavery will spread like a virus over the rest of the planet with predictable results. It's up to all that cherish human freedom and creativity to rise up, end this vile practice, and expose the global elite that secretly enable it. Only full disclosure of Antarctica's suppressed history and its current events will ensure that all humanity benefits from the advanced technologies that have been secretly developed and deployed there.

Disclosure will furthermore help prepare humanity for major geological events that lie ahead as west Antarctica's ice shelves continue to melt due to increased volcanic activity. The possibility that this not only causes a dramatic rise in sea levels, but also precipitates a geographical Pole Shift is something well worth considering. A global transformation awaits us if we have the courage to embrace the truth about what was, and is occurring now in Antarctica.

CHAPTER 1

The Enigmatic Thule Society:
Antigravity, Hitler & the German Navy

Thule Society Sponsors Antigravity Spacecraft R&D

The historic roots of the first secret space program and its genesis can be traced back to the early years of the Weimar Republic. At the end of the First World War in 1919, a number of German secret societies began collaborating in the development of flying saucer prototypes based on the designs received through the telepathic communications of an unusually beautiful and highly skilled psychic medium, Maria Orsic. When in a full trance state, she claimed to be in communication with a range of otherworldly beings. Among those were a group of Aryan or Nordic-looking extraterrestrials from the Aldebaran star system who wanted to assist humanity to develop spacecraft capable of interstellar flight. Orsic reported their intention in such an endeavor would raise human consciousness and accelerate humanity's evolution as a species in the galactic community.

Using automatic writing, Orsic wrote numerous pages of what appeared to be technical information in two foreign languages, though she did not recognize one language at all. Another prominent psychic, "Sigrun", assisted in understanding the content of the writing by getting clear mental images of a flying saucer craft. Orsic and Sigrun concluded that the information revealed how to build a spaceship.

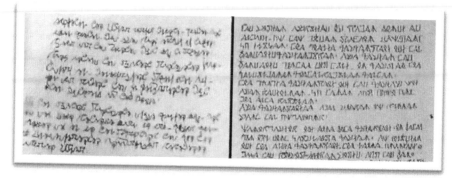

Figure 2. Automatic writing by Maria Orsic containing Sumerian and Templar languages.

Soon after, Orsic and Sigrun met with members from the Thule Society (Thule Gessellshaft), which after the First World War had become Germany's most powerful secret society, notably comprised of aristocrats and leading occultists of the day.[2] They were able to identify the second unknown language in the automatic writing as ancient Sumerian. Experts were brought in to translate Orsic's automatic writing. The translation confirmed Sigrun's mental images – it was in fact technical instructions for building a revolutionary type of engine that could power a spacecraft.[3] Orsic and supportive Thule society members arranged for various scientists to look at the translated information to determine whether it was scientifically feasible.

Orsic was enthusiastically supported by the Thule Society due to its member's belief in the existence of an advanced far northern (now subterranean) civilization described in Greek and Roman legends as Hyperborea, whose ancient capital was "Ultima Thule". In a preserved fragment by the Greek historian, Diodorus Siculus (1st century BC), he writes about this ancient civilization:

Now for our part, since we have seen fit to make mention of the regions of Asia which lie to the north, we feel that it will not be foreign to our purpose to discuss the legendary accounts of the Hyperboreans. Of those who have written about the ancient myths, Hecataeus and certain others say that in the regions beyond the land of the Celts there lies in the ocean an island no smaller than Sicily. This island, the account continues, is situated in the north and is inhabited by the Hyperboreans, who are called by that name because their home is beyond the point whence the north wind (Boreas) blows; and the island is both fertile and productive of every crop, and since it has an unusually temperate climate it produces two harvests each year... And the kings of this city and the supervisors of the sacred precinct are called Boreadae, since they are descendants of Boreas, and the succession to these positions is always kept in their family.[4]

The Boreadae were purported to be giant kings, around 10 feet tall by classical scholars.[5] Thule Society members believed that the Hyerboreans were the progenitors of the Aryan/Germanic race, and they exist to this present day, hidden away in the Earth's interior.

James and Suzanne Pool, authors of *Who Financed Hitler,* describe the powerful aristocratic members who financed and belonged to the Thule Society:

Outwardly, this mysterious group passed as a literary circle devoted to studying ancient German history and customs... The Munich branch had been financed during the war by a Baron Rudolf von Sebottednorff, a shadowy individual who enlisted over 250 members from the city and 1,500

3

throughout Bavaria. The significance of the membership, however, was not to be found in its quantity but its quality. Among the group's members were lawyers, judges, university professors, police officials, aristocratic members of the royal entourage of the Wittelsbachs, leading industrialists, surgeons, physicians, scientists, as well as rich businessmen like the proprietor of the elegant Four Seasons Hotel in Munich where the society had its headquarters.[6]

It is claimed that as early as 1917, Orsic met with Sebottendorf and three other Thule Society members in Vienna to discuss occult matters associated with her telepathic communications.[7] Based on the enthusiastic support she received, Orsic subsequently moved to Munich in 1919 to work with Sebottendorf and the Thule Society.

Thule Society leaders believed Orsic, and others like her, would offer important clues to understand a mysterious force called "Vril", a force that could be harnessed by individuals with sufficiently developed mental and psychic abilities. Such beliefs were very common among occult groups and secret societies in other nations as well, as exemplified in Edward Bulwer Lytton's 1871 book, *Vril: The Coming Race*.

The novel centers on a young, independently wealthy traveler (the narrator), who accidentally finds his way into a subterranean world occupied by beings who seem to resemble angels and call themselves Vril-ya. The hero soon discovers that the Vril-ya are descendants of an antediluvian civilization who live in networks of subterranean caverns linked by tunnels. It is a technologically supported Utopia, chief among their tools being the "all-permeating fluid" called "Vril", a latent source of energy that its spiritually elevated hosts are able to master through training of their will, to a degree which depends upon their hereditary constitution, giving them access to an extraordinary force that can be controlled at will. The powers of the will include the ability to heal, change, and destroy beings and things; the destructive powers in particular are awesomely powerful, allowing a few young Vril-ya children to wipe out entire cities if necessary. It is also suggested that the Vril-ya are fully telepathic.[8]

Significantly, Lytton also described the Vril-ya as possessing "air-boats" that could direct the Vril force as a destructive energy beam over a distance of 600 miles.

I should say, however, that this people have invented certain tubes by which the vril fluid can be conducted towards the object it is meant to destroy, throughout a distance almost indefinite; at least I put it modestly when I say from 500 to 600 miles. And their mathematical science as applied to such purpose is so nicely accurate, that on the report of some observer in an air-boat, any member of the vril department can estimate unerringly the nature of intervening obstacles, the height to which the projectile instrument should be raised, and the extent to which it should be charged, so as to reduce to ashes within a space of time too short for me to venture to specify it, a capital twice as vast as London.[9]

What Bulwer Lyton is referring to here in modern terms appears to be an aircraft equipped with a Directed Energy Weapon whose destructive capability exceeds the combined atomic bombs dropped on Hiroshima and Nagasaki!

Ominously, Lytton's book carried with it the warning that in the future, the surplus population of the subterranean Vril-ya civilization were destined to emerge on the Earth's surface. Through their advanced psychic powers and technologies, they would wipe out inferior or barbaric races in order to claim suitable territories. Those races which could sufficiently evolve in developing their inherent psychic abilities and technologies would become suitable partners for the emerging Vril-ya who would inevitably establish global domination.

While Lytton's book was published as a work of fiction, many occultists widely viewed it as factually based, and derived from ancient texts describing one or more hidden antediluvian civilizations. Lytton's status as a prominent member of the Rosicrucian Society, which has secretly studied ancient texts and encouraged members to develop higher consciousness and occult abilities, no doubt influenced widespread belief in the factual accuracy of *Vril: The Coming Race.*

Those who had developed such inherent psychic abilities would not only possess great power and rise to leadership positions, but would also be able to find and use incredibly advanced technologies hidden all over the planet. Some of these hidden technologies involved detailed designs of craft capable of space travel, which Thule Society members believed had first been developed by the ancient Hyberborean civilization. Importantly, these technologies were used by the hidden civilizations in the Earth's interior who were destined to emerge one day on the planet's surface.

In order to understand the role of the Thule Society in the creation of the Nazi Party and its evolution, a vital book to reference is *The Morning of the Magicians* by French authors, Louis Pauwels and Jacques Bergier. They point out in the beginning that Thule Society members believed that:

... not all the secrets of Thule had perished. Beings

intermediate between man and other intelligent beings from Beyond would place at the disposal of the Initiates a reservoir of forces which could be drawn on to enable Germany to dominate the world again and be the cradle of the race of Supermen which would result from mutations of the human species. One day her legions would set out to annihilate everything that had stood in the way of the spiritual destiny of the Earth, and their leaders would be men who knew everything, deriving their strength from the very foundation-head of energy and guided by the Great Ones of the Ancient World.[10]

Consequently, Thule Society members deemed it critical to establish an alliance with the Inner Earth (Hyperborean) beings who were destined to emerge in force upon the planet's surface:

Alliances could be formed with the Master of the World or the King of Fear who reigns over a city hidden somewhere in the East. Those who conclude a pact will change the surface of the Earth and endow the human adventure with a new meaning for many thousands of years. . . The world will change: the Lords will emerge from the center of the Earth. Unless we have made an alliance with them and become Lords ourselves, we shall find ourselves among the slaves, on the dungheap that will nourish the roots of the New Cities that will arise.[11]

This led to Pauwels and Bergier's pivotal conclusion that the Thule Society "took on its true character as a society of Initiates in communion with the *Invisible*, and became the magic center of the Nazi movement."[12] In short, they considered the Thule

Society to be the "secret directing agent of the Third Reich."[13]

The Thule Society enthusiastically supported Orsic and the group of young ladies she gathered around herself in Munich who were also psychically gifted. This is the period when Orsic created the "Alldeutsche Gesellschaft für Metaphysik" (Pan-German Society for Metaphysics) that was later renamed the Vril Society (Society of Vrilerinnen Women).[14] Another name change apparently occurred in 1941 when Hitler outlawed secret societies. Orsic registered the society as a business, named "Antriebstechnische Werkstätten" (Vril Propulsion Workshops).[15] These psychically gifted women dedicated themselves to developing techniques for communicating with otherworldly beings, and learned how to harness the Vril force for psychic purposes and space flight. Orsic herself was deeply dedicated to promoting greater awareness of humanity's latent spiritual potential and cosmic connections.

James and Suzanne Pool describe the importance of maintaining pure Germanic bloodlines for Thule Society members whose primary task was to rediscover the glories of the ancient Hyperborean civilization established by their distant ancestors:

> Only those who could prove their racial purity for at least three generations were admitted to this organization, whose motto was: Remember that you are a German! Keep your blood pure…. Like many other volkisch (racial, nationalist movements in Germany), the ostensible objective of the Thule Society was the establishment of a Pan-Germanic state of unsurpassed power and Grandeur.[16]

Other German secret societies like the "Die Herren vom schwarzen Stein" ("The Lords of the Black Stone") shared similar esoteric beliefs, and therefore, backed Orsic's exotic spacecraft development program. The necessary funding and scientific expertise were subsequently found to cultivate the first working prototypes based on the designs received by Orsic.

Professor Winfried Schumann, Director of the Electrophysical Laboratory from the Technical University of Munich (1924-1961), was a member of the Thule Society and was tasked to build Orsic's first craft.[17] Schumann was an expert in high energy plasma and high voltage electrostatics, both of which were key elements in the development of exotic propulsion systems for the proposed spacecraft.

It has been documented that Schumann investigated the free energy device of the German inventor, Hans Coler (aka Kohler). This device would eventually generate sufficient electric energy to power a submarine, and later a spacecraft. During the 1920's, the German Navy was actively researching new propulsion systems for its future generation of U-boats, which were being secretly assembled outside of Germany. In 1925, the German Navy had arranged for Schumann to evaluate the "Coler Device", despite the German patent office rejecting Coler's patent application. A report by a British Intelligence Sub-Committee cites Schuman's enthusiastic support of the Coler Device as an effective means of generating a "new source of energy":

> After the present examination, carried through as carefully as [possible], I must surmise that we have to face the exploitation of a new source of energy whose further developments can be of an immense importance. The apparatus was visible and accessible in all its essential parts. The inventor agreed quite willingly to each trial in so far as, according to his statement, no harm could be done to the working of the apparatus. I do not believe in a deception. I deem it expedient to put the apparatus to a further test, and I believe that a further development of the apparatus and an assistance, given to the inventor, will prove justified and of great importance.[18]

Figure 3. Hans Coler Free Energy Device

It's worth noting that three years after Schumann had tested the Coler device, U.S. inventor Thomas Townsend Brown was awarded a British patent for an electrogravitic device that was purported to have developed a new form of propulsion.[19] In a subsequent 1929 paper, Brown described how Einstein's efforts to develop a unified field theory had inspired him to find a fundamental connection between matter, gravity and electricity.

> There is a decided tendency in the physical sciences to unify the great basic laws and to relate, by a single structure or mechanism, such individual phenomena as gravitation, electrodynamics and even matter itself. It is found that matter and electricity are very closely related in structure. In the final analysis matter loses its traditional individuality and becomes merely an "electrical

condition." In fact, it might be said that the concrete body of the universe is nothing more than an assemblage of energy which, in itself, is quite intangible. Of course, it is self-evident that matter is connected with gravitation and it follows logically that electricity is likewise connected. These relations exist in the realm of pure energy and consequently are very basic in nature. In all reality *they constitute the true backbone of the universe*. It is needless to say that the relations are not simple, and full understanding of their concepts is complicated by the outstanding lack of information and research on the real nature of gravitation.[20]

It is certain that Schumann would have been aware of Townsend-Brown's device, and the radical new theories that underscored it. Combining the scientific breakthroughs behind the Coler device and the Townsend Brown electrogravitics device, Schumann had the scientific means for building the energy and propulsion systems for future German submarines, and more notably, spacecraft.

Documentary evidence further substantiating Schumann's involvement in German research and development of exotic propulsion systems is found in a 1946 US Army Air Force document that includes Schumann on a list of German scientists requested to work at Wright Field under Operation Paperclip.

W.O. Schumann

Second page of a three page declassified Operation Paperclip Memorandum dated 6 June 1947. This is a list of of German scientists requested by the U.S. Army Air Force for classified research at its Dayton, Ohio facilities. The appearance of Schumann's name is evidence that after his debriefing in post-war Germany, his expertise in aerospace projects was deemed important for the Army Air Force's classified foreign technology research. Source: Richard Sauder, *Hidden in Plain Sight* (2011).

Figure 4.   Professor Schumann shown to be part of Operation Paperclip.

Thule Society Chooses Hitler to lead a Pan-Germanic Workers Movement

The rise of Adolf Hitler was made possible by the Thule Society which secretly backed him. Prominent Thule Society members had sponsored the creation of the German Workers Party espousing nationalist ideals in order to prevent the working class from falling under the influence of the rapidly growing Communist movement.

> Unlike most other conservative nationalists, the Thule Society was aware of the dangers presented by the widening gap between the officer class and the workers. It became one of the society's primary objectives to bring the working man back into the national camp…. Given the existing sentiments of class hostility the Thule program would be automatically rejected by the masses if proposed by someone of a privileged class.[21]

The creation of the German Worker's Party took place on January 5, 1919, and the Thule Society initially supported Anton Drexler to lead it as a figurehead who they would use and secretly control in order to achieve the Society's pan-Germanic, anti-Semitic and metaphysical agenda.[22]

Drexler, however, lacked the organizational, oratorical and charismatic skills necessary to build up membership for a large workers party espousing a pan-Germanic agenda. Fatefully, the necessary leader was found when a young German Army spy attended one of the German Workers Party's meetings, as the late Jim Marrs writes in his book, *Rise of the Fourth Reich*:

> It was in this setting that Hitler, a twenty-nine-year- old veteran, came into contact with members

of the "Thule Gesellschaft", or Thule Society, ostensibly an innocent reading group dedicated to the study and promotion of older German literature. But the society, composed mostly of wealthy conservatives, ardent nationalists, and anti-Semites, actually delved into radical politics, race mysticism, and the occult under its emblem— a swastika superimposed over a sword. The society also served as a front for the even more secretive Germanenorden, or German Order, a reincarnation of the old Teutonic Knights, which had branches throughout Germany patterned after Masonic lodges. It is believed that these lodges carried on the agenda of the outlawed Bavarian Illuminati, with its fundamental maxim that "the end justifies the means."[23]

Marrs' view that the Thule Society was a new incarnation of the banned Bavarian Illuminati is important to keep in mind given its behind-the-scenes role in Hitler's rise to power and the emergence of the Third Reich.

James and Suzanne Pool assert that Hitler would have quickly learned that the German Workers Party was sponsored by the Thule Society, and had agreed to work with prominent Thule members in achieving their commonly shared pan-Germanic, anti-Semitic and metaphysical beliefs:

[W]as Hitler aware at the time he joined that the German Workers Party was backed by the Thule Society? ... Considering Hitler's position as an Army agent and his interest in nationalistic anti-Semitic politics, it is likely that he was aware of the society's backing this new little movement called the German Workers Party. If Hitler had such information it would explain why he chose this small party from the many other nationalist groups which were in existence at the time. [24]

Thus, the Thule Society began supporting the German Workers Party, which in early 1920, at Hitler's insistence, changed its name to the National Socialist German Workers Party or Nazi Party.[25] The Thule Society organized for its Munich based newspaper, the *Volkischer Beobachter* (*National Observer*), "the leading right-wing anti-Semitic paper in Bavaria", to promote the new Party's agenda.[26] The newspaper was soon handed over to the Nazi Party, which had mysteriously raised the funds to purchase it, and it henceforth became the official mouthpiece of the Nazi Party. This was a critical step in transforming the fledgling Nazi Party into a mass movement, and for introducing Hitler to a national audience. All of this was achieved through a newspaper that was first established and secretly controlled by prominent Thule Society members:

> When the Thule Society turned the *Volkischer Beobachter* over to the Nazi Party, it must have been specified in the deal that a Thule member (Amann) would remain in charge of the newspaper's funds, and moreover, would be appointed 'Party business manager' with control over all Party money.... With Amann as Party business manager, Eckart as editor of the Party newspaper, and Rosenberg as assistant editor, the Thule Society's involvement with the Nazis was stronger than ever. But since the basic ideology of the Thule Society and the Nazi Party were the same, these men could be loyal Nazis as well as members of the society.[27]

It is not just vital organizational support from Thule Society members that proved essential for Hitler's rise to power. It was also the ideas and motifs of the Thule Society which Hitler used to appeal to a wide pan-Germanic base, according to the Pool's:

> The symbol of the Thule Society was the swastika. Its letterheads and literature displayed the emblem, and large swastika flags decorated its plush meeting rooms and offices. Many of the themes and slogans of the group were later repeated by Hitler almost word for word.[28]

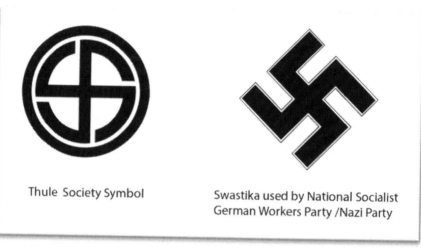

Thule Society Symbol

Swastika used by National Socialist German Workers Party /Nazi Party

Figure 5. Thule Society use of Swastika adapted by the Nazi Party.

The Swastika represented the coveted Vril force and thus was a vital aspect of what the Thule Society believed all pure blooded Germans had to learn and master. Hitler's adoption of many Thule Society ideas and motifs is evidence that he was himself an occultist, and even a Thule member as claimed by historical researchers such as Trevor Ravenscroft, author of *The Spear of Destiny*.[29]

In addition, Thule Society members provided vital support to Hitler in the face of police prosecution for his outspoken nationalist views. As the Pool's write:

The most significant assistance he would get from the Thule Society for the time being was protection from police prosecution thanks to Thule members in the Bavarian government. [30]

According to the Pool's, Hitler's fledgling Nazi party would eventually outgrow its Thule Society patron:

[A]s the German Workers Party would begin to grow and develop under Hitler's guidance, able and intelligent Thule sympathizers and members would join it and be of the utmost value to Hitler. Eventually the child of the masses would outgrow its secret society parent. [31]

Historians largely agree that the Thule Society backed Hitler's rise to power, but like the Pool's, dismiss its long term significance given its apparent dissolution in 1925, the same year Hitler's autobiography *Mein Kampf* (My Struggle) was published, notably with Thule Society support. Similarly, Nicholas Goodrick-Clark, author of *The Occult Roots of Nazism* writes: "The Thule Society was dissolved around 1925 when support had dwindled." [32]

This is where the Pool's and Goodrick-Clark fail to take into account the Thule Society's behind-the-scenes role as an occult group, not a political lobby group. Rather than disappearing from the national scene and having its members absorbed by the growing Nazi Party, the Thule Society succeeded in infiltrating Hitler's Nazi Party to place prominent members in key positions to achieve its long term goals. As cited earlier, Pauwels and Bergier saw the Thule Society as the "magic center of the Nazi movement". [33] Many additional sources point to the Thule Society's continued existence behind the scenes as an underground, but highly powerful, organization that silently worked with and manipulated Hitler long after its apparent public dissolution in 1925.[34]

As a modern day incarnation of the banned Bavarian Illuminati, Thule Society members were very familiar with exerting influence from behind the scenes in order not to generate alarm over their hidden influence. After all, the Bavarian Illuminati had been outlawed once before due to the outcry by an alarmed public, resulting in the seizure of its property and wealth by the Bavarian government. Hitler shared the pan-Germanic, anti-Semitic and metaphysical goals of the Thule Society, so why jeopardize them by revealing the Thule Society's true behind-the-scenes role? Hitler was very eager to show his independence and not expose himself as an occult practitioner or follower of an aristocratically led group such as the Thule Society.

According to Ian Kershaw, author of *Hitler: 1889-1936*, the Thule Society's "membership list... reads like a Who's Who of early Nazi sympathizers and leading figures in Munich", which included Rudolf Hess, Alfred Rosenberg, Hans Frank, Julius Lehmann, Gottfried Feder, Dietrich Eckart, and Karl Harrer.[35] An even more expansive list was provided by the authors Dietrich Bronder (*Before Hitler Came*) and E.R. Carmin (*Guru Hitler*), who also enumerated many Thule Society members, some of whom rose to become powerful officials in the Nazi Government:

- Baron Rudolf von Sebottendorf, Grand Master of the Order
- Guido von List, Master of the Order
- Jorg Lanz von Liebenfels, Master of the Order
- Adolf Hitler, "Fuhrer", German Chancellor, SS Superior
- Rudolf Hess, Vice Fuhrer, and SS Obergruppenfuhrer
- Herman Goering, Reichsmarschall and SS Obergruppenfuhrer

- Heinrich Himmler, Reichsfuhrer SS and Reichsminister
- Alfred Rosenberg, Reichsminister and NS-Reichsleiter
- Hans Franck, Dr. Dr. h.c., NS-Reichsleiter and Governor General of Poland
- Julius Streicher, SA-Obergruppenfuhrer and Gauleiter of Franken
- Karl Haushofer, Prof. Dr., Major General ret.
- Gottfried Feder, Prof. Dr., Secretary of State ret.
- Dietrich Eckart, Editor in Chief of the Volkischer Beobachter
- Bernhard Stempfle, father confessor and confidant of Hitler
- Theo Morell, personal physician of Hitler
- Franz Gurtner, president of the police, Munich
- Rudolf Steiner, founder of the anthroposophic teaching
- W.O. Schumann, Prof. Dr. at the Technical University Munich
- Trebisch-Lincoln, occultist and traveler to the Himalayas
- Countess Westrap
- AND MANY OTHERS...[36]

Among Thule Society members, Dietrich Eckart was of particular importance to Hitler early in his political career by providing access to powerful people willing to fund the Nazi party. The Pool's write:

> In light of his success, more important members of the Thule Society began to join the German Workers Party... One of them Dietric Eckart came

to exert a tremendously powerful personal influence on Hitler.... He gave Hitler his first introduction to better society and, what is more important, to people who were financial backers of the Thule Society.[37]

It was Eckert who made it possible for the Nazi Party to raise the necessary funds to purchase the *Volkischer Beobachter* newspaper, and he became its chief editor up to his death in 1923.[38] Hitler dedicated the second volume of *Mein Kampf* to Eckart, illustrating the powerful influence this leading Thule Society member had upon him.

Another strong influence on Hitler was the retired Army General and then University of Munich Professor, Karl Haushofer. His pioneering research in Japan and Asia led to many insights about ancient history, the role of the Aryan race and understanding the mysterious Vril force. Haushofer was also a member of the Thule Society and passed on his teachings to an enthusiastic student and Thule member, Rudolf Hess, who became Hitler's right hand man.

It was Hess who helped Hitler in the writing of *Mein Kampf*, which contained many of Haushofer's seminal ideas.[39] In 1945, after the defeat of Nazi Germany, Haushofer explained to his US Army interrogators, "I was only able to influence [Hitler] through Hess."[40] This was a very telling admission. Hess in reality was Hitler's handler on behalf of the Thule Society, a role he would continue to play right up to 1941, when he was captured during his failed peace mission to Scotland. Hess was part of an orchestrated effort to reach a peace deal with a pro-German faction in the British aristocracy, which included prominent members of the Royal Family – many of whom had strong ties to Thule and other Secret Society members within the Nazi Government.

German Naval Intelligence Support of Thule Society & Hitler

The Thule Society had a critical institutional ally in promoting its pan-Germanic, anti-Semitic and metaphysical agenda during the years of the Weimar Republic – the German Navy (Reichsmarine, renamed Kriegsmarine during WWII). From its inception, the Thule Society had been supported by the German Navy due to its staunch opposition to the Treaty of Versailles, advocacy of German national unity, and its commitment to rapid rearmament. The German Navy and other German nationalist organizations recognized that the Versailles Treaty had the effect of encouraging secessionist movements due to the hardships it imposed, and it conveyed the idea that breakaway states would be exempted from its onerous provisions. A breakup of Germany would remove it as a potential geo-political rival to Britain and France for decades, and make economic revival that much harder.

Towards the end of World War I, Bavaria had a powerful secessionist movement, which was supported by the Communist movement. From April 6 to May 3, 1919, a Bavarian Soviet Republic was briefly in power, but after the execution of eight aristocratic members of the Thule Society, the Republic was violently suppressed. Bavarian secessionism was vehemently opposed by the Thule Society and the German Navy. Thus, cooperation was natural and inevitable between these organizations, and quickly manifested once the Thule Society was formed in early 1919. The first evidence for this reveals itself in the Navy providing space for Thule meetings, as the Pool's note:

> Before the Thule Society rented their own offices in the Four Seasons Hotel, they held their meetings in the rooms of the Naval Officers Club in the same hotel. The Thule Society supported and later

sheltered many of the officers and men of the Ehrardt Brigade, a naval unit, after the Kapp putsch failed. [41]

Yet another powerful reason for the German Navy's support for the Thule Society comes out of the Treaty of Versailles, which severely limited the number of ships the Navy could own, as well as abolishing entire fleet categories such as submarines. Article 181 of the treaty stipulated:

> After the expiration of a period of two months from the coming into force of the present Treaty the German naval forces in commission must not exceed: 6 battleships of the Deutschland or Lothringen type, 6 light cruisers, 12 destroyers, 12 torpedo boats, or an equal number of ships constructed to replace them as provided in Article 190.
>
> No submarines are to be included.
>
> All other warships, except where there is provision to the contrary in the present Treaty, must be placed in reserve or devoted to commercial purposes. [42]

In order to circumvent the Treaty restrictions, Naval Intelligence had arranged to work with major German companies to secretly build naval assets in foreign nations, until such time as the restrictions were lifted and rearmament could openly proceed. To this end, German Naval Intelligence used the Thule Society to funnel financial support to Hitler, and was an important covert funder of the Nazis in their rise to political power.

> Besides the aristocrats, businessmen, and White Russians, there remained one other group from

which Hitler received money: Naval Intelligence....
Involved in this case was the man who was later to
become known as a leader of the German
resistance against Hitler (then Lieutenant
Commander) Wilhelm Canaris. Since intelligence
agencies are expert at covering up any traces of
their activities, the evidence that remains is rather
sketchy. However, it is enough to definitely link
Hitler, the Organization Consul (a right-wing
terrorist unit of former naval officers), and funds of
Naval Intelligence.[43]

Wilhelm Canaris, who, if not an outright member, at least shared
many Thule Society beliefs, not only helped to secretly fund Hitler
with Naval Intelligence funds, but also assisted in organizing the
Nazi paramilitary units that were instrumental in street fighting
against communists and other radical leftist groups. [44] Canaris'
role in helping fund and organize the Nazis was later rewarded by
Hitler who had him appointed chief of German military
intelligence (Abwehr) from 1935 to 1944.

Soon after the end of World War I hostilities, German
Naval Intelligence began working with major steel companies such
as Thyssen, Frick, Krupp, and the giant chemical conglomerate,
I.G. Farben (formed in 1925), to secretly direct funds to where
they needed to be for the covert Naval rearmament program.
Countries such as Spain and Holland would become important
intermediaries in building U-boats (submarines), which would
eventually be reassembled in Germany. Ian Colvin, author of a
biography on Canaris, wrote:

> The German Navy would be expanded, the U-boats
> that Canaris had hitherto helped to lay down
> secretly in Spain and Holland would be assembled
> at Hamburg, Bremen, and the Baltic ports after
> being prefabricated at inland factories.[45]

Later, I will examine how the knowledge and connections Canaris acquired in the covert building of submarine components and other military assets – in foreign lands using German companies, and their international partners and subsidiaries during the Weimar Republic years (1919-1933) – would prove invaluable over a decade later when bases were slated to be built in Antarctica after the first Nazi expedition there in early 1939.

After becoming head of Nazi Germany's military intelligence in 1935, Canaris soon began orchestrating a covert armament program outside of Germany, once again using leading German companies and their foreign partners. However, this time the futuristic project would involve building fleets of antigravity spacecraft; using advanced technologies discovered by Thule/Nazi scientists around the world, along with the paranormal communications, and later, actual contact with different extraterrestrial races.

Lastly, coming back to the period immediately following World War I, a critical component stands out in the covert rearmament program spearheaded by the German Navy; foreign banks. These savvy institutions could finance such military construction endeavors outside of Germany without any suspicion being raised over the ultimate destination of the weapons related products being built. August Thyssen, who owned one of Germany's largest steel firms, Thyssen AG (founded in 1891) had opened a number of banks to facilitate business interests and contracts with the German military. Thyssen's foreign banks would play a key role as the intermediary in paying international contractors for the German Navy's covert rearmament program. Notably, it was U.S. banks and corporations that played a significant role in providing funds for August Thyssen's banks and the German Naval Intelligence slush fund they were associated with.

CHAPTER 2

Role of U.S. Corporations
in Nazi Germany

U.S. Corporations Aid Covert Rearmament of German Navy & Nazi Party

As World War I drew to a tumultuous close in 1918, August Thyssen opened the Bank Voor Handel en Scheepvaart, N.V. in Rotterdam, the Netherlands.[46] It was a subsidiary of his Berlin based August Thyssen Bank of Germany, and was created to be his own personal banking operation. The Dutch bank would provide a safeguard for Thyssen and his two sons to cleverly move significant funds at a moment's notice, depending on the outcome of the First World War. This became particularly important as Germany lurched into a chaotic period during the final stages of the war, that led up to the signing of an armistice ending hostilities on November 11, 1918. The Treaty of Versailles was signed by the defeated Germans over seven months later on June 28, 1919, and many hardships laid ahead as German industry attempted recovery in the midst of onerous reparation payments that led to little in the way of government contracts.

By placing funds in its Dutch bank, the Thyssen's company was able to protect its remaining assets, recover from the devastating war losses, rebuild its steel business, support anti-communist nationalist movements, and finally play a key role in the covert rearmament of the Germany military, particularly the Navy. The Thyssen family wholeheartedly supported the Thule Society and the German Navy's effort to promote a strong nationalist movement. Hitler's burgeoning Nazi Party was the most promising of all the nationalist parties, and the Thyssen's saw opportunity in financially supporting it.

Now, August Thyssen's two sons would play diametrically different political roles in order for the family to protect itself in the future from the kind of losses experienced from the First World War. According to John Loftus, a former U.S. Department of Justice Nazi War Crimes prosecutor:

> After World War I, August Thyssen had been badly burned by the loss of assets under the harsh terms of the Versailles treaty. He was determined that it would never happen again. One of his sons would join the Nazis; the other would be neutral. No matter who won the next war, the Thyssen family would survive with their industrial empire intact. Fritz Thyssen joined the Nazis in 1923; his younger brother married into Hungarian nobility and changed his name to Baron Thyssen-Bornemisza. The Baron later claimed Hungarian as well as Dutch citizenship. In public, he pretended to detest his Nazi brother, but in private they met at secret board meetings in Germany to coordinate their operations. If one brother were threatened with loss of property, he would transfer his holdings to the other.[47]

Fritz Thyssen is "remembered as the man who gave more money to Hitler than any other individual."[48] His first donation of 100,000 gold marks ($25,000) was given to the Nazi Party after he first heard Hitler speak in 1923.[49] Thyssen described how he was persuaded by World War I General Erich Ludendorff, former Quartermaster general (the second most powerful military official in all of Germany) to attend the Hitler speech:

> "There is but one hope," Ludendorff said to me, "and this hope is embodied in the national groups which desire our recovery." He recommended to me in particular the Overland League and, above all, the National Socialist party of Adolf Hitler. Ludendorff greatly admired Hitler. "He is the only man," he said, "who has any political sense. Go and listen to him one day." I followed his advice. I attended several public meetings organized by Hitler. It was then that I realized his oratorical gifts and his ability to lead the masses. What impressed me most, however, was the order that reigned in his meetings, the almost military discipline of his followers.[50]

Ludendorff understood the important role nationalist movements could play in Germany's covert rearmament program and overturning the Versailles Treaty.

International investors were needed to raise funds for the Thyssen family's multiple projects, which would come to include the covert funding of Hitler's Nazi Party. In 1922, when U.S. banker Averell Harriman travelled to Germany and met with the Thyssen family, a bargain was struck that investment opportunities could be best facilitated by jointly creating a private bank in New York; the Union Banking Corporation, which was established in 1924. Harriman's partner, George Herbert Walker (grandfather of President George Herbert Walker Bush) became the President of the new bank. In 1926, he appointed his son-in-law, Prescott Bush, to be the Vice-President of the Union Banking Corporation. The newly established bank "made it easy for the

Thyssens to move assets around, launder money, conceal profits and evade taxes."[51] In turn, the Union Banking Corporation helped Prescott Bush amass a fortune which he used effectively in his campaign to get elected to the U.S. Senate, and this wealth enabled him to establish a political dynasty paving the way for a son (George H.W. Bush) and grandson (George W. Bush) to become future U.S. Presidents.

Despite the Union Banking Corporation having American directors, its principal owners were the Thyssen family through its Dutch bank subsidiary, as an August 16, 1941 investigation into the Union Banking Corporation confirmed:

> "Union Banking Corporation, incorporated August 4 1924, is wholly owned by the Bank voor Handel en Scheepvaart N.V of Rotterdam, the Netherlands. My investigation has produced no evidence as to the ownership of the Dutch bank. Mr Cornelis [sic] Lievense, president of UBC, claims no knowledge as to the ownership of the Bank voor Handel but believes it possible that Baron Heinrich Thyssen, brother of Fritz Thyssen, may own a substantial interest."[52]

Investigators furthermore concluded that: "the Union Banking Corporation has since its inception handled funds chiefly supplied to it through the Dutch bank by the Thyssen interests for American investment."[53] Webster Tarpley, author of *George Bush: The Unauthorized Biography,* wrote:

> Thus by personal agreement between Averell Harriman and Fritz Thyssen in 1922, W.A. Harriman & Co. (alias Union Banking Corporation) would be transferring funds back and forth between New York and the "Thyssen interests" in Germany. By putting up about $400,000, the Harriman

organization would be joint owner and manager of Thyssen's banking operations outside of Germany...[54]

In 1926, Fritz Thyssen took over his father's steel company, and two years later he formed a corporate conglomerate, United Steelworks (Vereinigte Stahlwerke AG), which controlled more than 75% of Germany's iron ore reserves. Thyssen's conglomerate, which included the newly renamed company from his father, August Thyssen-Hutte AG, became essential to steel production for Nazi Germany up to and throughout World War II, and for its covert naval rearmament program using foreign banks and subsidiaries in the Netherlands, U.S. and elsewhere around the world.

Other German industrialists collaborated with the Thyssens in using foreign subsidiaries and banks to move funds and participate in the covert rearmament program spearheaded by the German Navy. Among them was Friederich Flick, who partnered with Thyssen in setting up United Steelworks, and worked through the Union Banking Company, as Webster Tarpley describes:

> *Friedrich Flick* was the major co-owner of the German Steel Trust with Fritz Thyssen, Thyssen's long-time collaborator and occasional competitor.... The Flick-Harriman partnership was directly supervised by Prescott Bush, President Bush's father, and by George Walker, President Bush's grandfather.[55]

In 1926, Allen Dulles (the future Director of the CIA) joined his brother, John Foster Dulles (future Secretary of State under President Eisenhower) at the prestigious law firm, Sullivan and Cromwell, which John Foster was managing at the time. James Srode, author of a biography of Allen Dulles, described the company's influence at the time:

> It was upon Foster Dulles' accession to the post of managing partner, the biggest, most powerful, most respected law firm in the world. Its clients were national governments, major corporations, even entire industries.[56]

Among the Dulles brother's many clients was the same Dutch bank belonging to the Thyssen family, Bank voor Handel en Scheepvaart N.V of Rotterdam. The Dulles brothers also represented a number of other German companies, including the giant I.G. Farben chemical conglomerate, which along with the Thyssens and Flicks' United Steelworks, would play an essential role in the rearmament of Germany and the development of its fledgling secret space programs, one of which would be set up in Antarctica.

U.S. Corporations and Hitler's Rise to Power

Hitler was chosen to be the charismatic figurehead for a pan-Germanic nationalist movement based on the ideas of territorial expansion and racial supremacy as he most clearly enunciated in his 1925 book, *Mein Kempf*. Underlying these pan-Germanic views was the Thule Society belief that the German race was directly descended from the Hyboborean's and their great civilization, and thus the most likely to succeed in finding and developing the latter's hidden ancient technologies using the Vril force, symbolized by the Swastika.

The Nazi Party was vehemently anti-communist, hostile to Jewish influence, and opposed to the growing power of the labor movement. Hitler's Nazi Party, now fully infiltrated by Thule Society members, aimed to work closely with German industrialists led by the Thyssen family, who wanted to restore Germany's manufacturing potential through favorable Government policies. In 1933, Thyssen led the way in organizing a letter by 39 German industrialists which included famous business names such as Krupp, Siemens and Bosch to call for the aging President Paul Von Hindenberg to appoint Hitler as Chancellor of Germany – a position similar to Prime Minister in the Westminster system of Government.[57] This led to a meeting on January 4, 1933 where a deal was struck between Hindenberg and a group of German aristocrats, industrialists and army officers.[58]

On January 30, Hitler became Chancellor, albeit with minority party support, in the Reichstag (German Parliament). Hitler immediately called for new elections to be held on March 5, 1933, aiming to increase the number of Nazi Party representatives in the Reichstag (only 196 out of a total 647), which was critical to his plans to expand his power as Chancellor. Thyssen once again was instrumental in supporting Hitler and arranged for the Association of German Industrialists to donate three million Reichmarks to the Nazi Party for the March election.

Hitler was successful in increasing the number of Nazi Party representatives from 196 to 288. Together with the 52 seats of its coalition partner, the German National People's Party, the Nazis had achieved a staunch majority in the Reichstag. This opened the pathway for Hitler's power as Chancellor to swell. In thanks, Hitler arranged for Thyssen to be elected a Nazi member of the Reichstag and appointed him to the Council of State of Prussia, both of which were honorary positions. Thyssen's influence and his role in bringing Hitler to power was immortalized in the August 1933 edition of the German magazine, AIZ, showing Thyssen pulling Hitler's strings on behalf of German industry. The German caption reads: "Tool in god's hands. Toy in Thyssen's hands."

Figure 6. German Magazine AIZ illustrates Fritz Thyssen's influence over Adolf Hitler in 1933.

After having consolidated his power, Hitler quickly began cracking down on Communists, the labor movement and the Jewish population. Policies were passed which enabled German Corporations to utilize Germany's highly skilled population at bargain basement rates. German manufacturing took off with abundantly cheap labor where workers were prevented by the government from any attempt to improve worker conditions through industrial action against their corporate masters. Thyssen and other major German industrialists were amply rewarded for supporting Hitler's rise to power. Most significantly, Hitler's government was very favorable to international corporations that wanted to partner with German companies in building large factories that utilized Germany's cheap labor force. Once again, Thyssen played a key role in this process.

According to author and political writer, Eustace Mullens, both Allen and John Foster Dulles attended the January 4, 1933 meeting between Hindenberg and German industrialists resulting in pledged support for Hitler since he had promised to break the power of the trade union movement if he became Chancellor.[59] Most historians dispute the Dulles brothers' attendance at the meeting. What is known without dispute, however, is that both Dulles brothers did meet with Hitler in Berlin in the months of April and May 1933, and that the German companies the brothers represented, through Sullivan and Cromwell, were part of the industrialist group that brought Hitler to power.

James Srodes, one of Allen Dulles biographers, confirms that Dulles met with Hitler in April 1933, only a month after the elections that cemented his power within the Reichstag.[60] John Foster Dulles also attended the meeting with Hitler, which was part of a series of negotiations that culminated in agreements reached in May 1933 between their law firm, Sullivan and Cromwell, and dozens of German companies and Nazi clients, as Webster Tarpley explains:

In May 1933, just after the Hitler regime was consolidated, an agreement was reached in Berlin for the coordination of all Nazi commerce with the U.S.A. The *Harriman International Co.,* led by Averell Harriman's first cousin Oliver, was to head a syndicate of 150 firms and individuals, to conduct *all exports from Hitler Germany to the United States.*

This pact had been negotiated in Berlin between Hitler's economics minister, Hjalmar Schacht, and John Foster Dulles, international attorney for dozens of Nazi enterprises, with the counsel of Max Warburg and Kurt von Schroeder.

Throughout the 1930's, John Foster Dulles arranged debt restructuring for German firms under a series of decrees issued by Adolf Hitler. In these deals, Dulles struck a balance between the interest owed to selected, larger investors, and the needs of the growing Nazi war-making apparatus for producing tanks, poison gas, etc.[61]

Whether or not the Dulles brothers attended the secret January 1933 meeting between President Hindenberg and German industrialists, it is certain they would have been aware of it, and even encouraged it in order to promote the stability of the German economy. As Srodes, writes:

[Allen Dulles] recognized that Sullivan & Cromwell had a vested interest in preserving the stability of the German economy no matter who was in power; fully one-third of all the foreign bonds that defaulted to their American investors during the Depression were on loans made to German government units and businesses, bonds which the firm, and he and Foster personally, had a large hand in facilitating.[62]

Since Hitler was good for German businesses, then Hitler was good for the Dulles brothers and the investments of U.S. corporations such as Prescott Bush's Union Banking Corporation in Germany.

Among the major U.S. corporations that invested heavily in Nazi Germany was the Ford Motor Company. In fact, Henry Ford was cited by Adolf Hitler in *Mein Kampf* on the issue of how to deal with the Jewish question:

> ... only a single great man, Ford, [who], to [the Jews'] fury, still maintains full independence ... [from] the controlling masters of the producers in a nation of one hundred and twenty millions.[63]

In 1920, Ford had arranged for his personal newspaper, the *Dearborn Independent*, to release 91 stories about the "Jewish menace" which were published and distributed in a four volume set, called *The International Jew*.[64] Famously, in a 1931 interview with a Detroit News Reporter, "Hitler said he regarded Ford as his "inspiration," explaining his reason for keeping Ford's life-size portrait next to his desk."[65] Jim Marrs explains how Hitler's admiration for Ford was fully reciprocated:

> Ford became an admirer of Hitler, provided funds for the Nazis, and, in 1938, became the First American to receive the highest honor possible for a non-German: the Grand Cross of the Supreme Order of the German Eagle.[66]

The Rockefeller family-led oil giant, Standard Oil, also invested heavily in Germany through its partnership with the I.G. Farben conglomerate:

In 1934 Germany produced only 300,000 tons of natural petroleum products and synthetic gasoline. In 1944, thanks to the transfer of hydrogenation technology from Standard Oil of New Jersey to I. G. Farben, Germany produced 6,500,000 tons of oil, 85 percent of which was synthetic.[67]

Significantly, according to Paul Manning, a CBS news correspondent during World War II, Hermann Schmitz, the President of I.G. Farben, once held as much stock in Standard Oil of New Jersey as did the Rockefellers."[68]

Other leading U.S. corporations such as General Motors, IBM (International Business Machines) and ITT (International Telephone and Telegraph Corporation) established subsidiaries in Nazi Germany. U.S. Banks such as J.P. Morgan's First National Bank of New York, the Rockefeller's Chase National Bank and the National City Bank of New York backed the U.S. corporate investments in Germany, playing a major role in the rapid growth in German manufacturing industries and the military re-armament of Germany.[69]

Charles Higham, author of *Trading with the Enemy*, describes how another prominent U.S. Bank that directly supported Nazi Germany involved the Dulles brothers, and even had Allen Dulles on its Board of Directors:

> [I]n 1936, the J. Henry Schroeder Bank of New York had entered into a partnership with the Rockefellers. Named Schroeder, Rockefeller and Company, Investment Bankers, the firm became what Time magazine called the economic booster of "the Rome-Berlin Axis." "Avery Rockefeller owned 42 percent of Schroeder," Higham reported. "Their lawyers were John Foster Dulles and Allen Dulles of Sullivan and Cromwell. Allen Dulles (later of the Office of Strategic Services) was on the

board of Schroeder."[70]

In 1937, U.S. Ambassador to Germany William E. Dodd explained how close America's corporate elite was to Nazi Germany:

> A clique of U.S. industrialists is hell-bent to bring a fascist state to supplant our democratic government and is working closely with the fascist regime in Germany and Italy. I have had plenty of opportunity in my post in Berlin to witness how close some of our American ruling families are to the Nazi regime.[71]

Higham describes this transnational clique of industrialists and national elites that shared Fascist/Nazi ideals as "The Fraternity".[72] The Fraternity included secret societies that shared belief systems which transcended national affiliations. Thule Society ideas about an ancient Hyperborean race from which the Aryan race descended, and the importance of maintaining racial purity and mastering the Vril force were shared by The Fraternity, which today is called by other appellations, such as; the Cabal, Illuminati, the 13 ruling blood-line Families, etc.

Industrialists and elites formed a close alignment of interests across international borders, ignoring democratic ideals and laws passed by representative democracies. Even during the time of war, The Fraternity maintained their mutual trade despite legislation, making a mockery of national parliaments and rulings such as the *Trading with the Enemy Act*, which had been passed in the U.S. in 1917. Despite similar legislation in other countries, The Fraternity could move profits across borders to maximize earnings from both sides, and protect their investments by national appropriation.[73] U.S. companies were able to legally do this through a little known general license, authorized by President Roosevelt, that could be granted through the U.S. Treasury to companies, thereby, allowing them to circumvent the *Trading With the Enemy Act.*

Higham describes the extent to which U.S. corporations worked with their German partners not only leading up to World War II, but also unconscionably throughout it as well, due to the licenses granted to them:

> To this day the bulk of Americans do not suspect The Fraternity. The government smothered everything, during and even (inexcusably) after the war. What would have happened if millions of American and British people, struggling with coupons and lines at the gas stations, had learned that in 1942 Standard Oil of New Jersey managers shipped the enemy's fuel through neutral Switzerland and that the enemy was shipping Allied fuel? Suppose the public had discovered that the Chase Bank in Nazi occupied Paris, after Pearl Harbor was doing millions of dollars' worth of business with the enemy with the full knowledge of the head office in Manhattan? Or that Ford trucks were being built for the German occupation troops in France with authorization from Dearborn, Michigan? Or that Colonel Sosthenes Behn, the head of the international American telephone conglomerate ITT, flew from New York to Madrid to Berne, during the war to help improve Hitler's communications systems and improve the robot bombs that devastated London? Or that ITT built the Focke-Wulfs that dropped bombs on British and American troops? Or that crucial ball bearings were shipped to Nazi associated customers in Latin America with the collusion of the vice chairman of the U.S. War Production Board in partnership with Goring's cousin in Philadelphia when American

forces were desperately short of them? Or that such arrangements were known about in Washington and either sanctioned or deliberately ignored?[74]

Roosevelt had passed his presidential edict authorizing licenses to circumvent the *Trading with the Enemy Act* only six days after the December 7, 1941 Pearl Harbor attack.[75]

GENERAL LICENSE UNDER SECTION 3(a)
OF THE TRADING WITH THE ENEMY ACT

By virtue of and pursuant to the authority vested in me by sections 3 and 5 of The Trading with the Enemy Act as amended, and by virtue of all other authority vested in me, I, Franklin D. Roosevelt, President of the United States of America, do prescribe the following:

A general license is hereby granted, licensing any transaction or act proscribed by section 3(a) of The Trading with the Enemy Act, as amended, provided, however, that such transaction or act is authorized by the Secretary of the Treasury by means of regulations, rulings, instructions, licenses or otherwise, pursuant to the Executive Order No. 8389, as amended.

FRANKLIN D. ROOSEVELT

THE WHITE HOUSE,
December 13, 1941

H. MORGENTHAU, JR.
Secretary of the Treasury
FRANCIS BIDDLE
Attorney General of the United States

Figure 7. General Licenses could be granted to Companies trading with Nazi Germany during WWII.

Roosevelt's edict, according to Higham, was done to benefit the interests of The Fraternity, and served no genuine national security purpose. However, there is another plausible explanation to consider for why such licenses were granted. They would invaluably allow US Navy and Army spies to infiltrate key transnational corporations in Nazi Germany to gather intelligence on their advanced technology programs, just as William Tompkins claims regarding the intelligence program he participated in at Naval Air Station, San Diego, from 1942 to 1946. In a personal interview, Tompkins confirmed that U.S. companies were indeed used as a means of infiltrating the Nazis aerospace industries.[76]

In the case of the International Telegraph and Telephone company (ITT), which worked closely with U.S. military intelligence, it was among the U.S. companies granted licenses to continue working with Axis powers right up to 1945, as Higham writes:

> ITT was allowed to continue its relations with the Axis and Japan until 1945, even though this conglomerate was regarded as an official instrument of United States Intelligence... [I]n the case of ITT, perhaps the most flagrant of the corporations in its outright dealings with the enemy, Hitler and his postmaster general, the venerable Wilhelm Ohnesorge, strove to impound the German end of the business. But even they were powerless in such a situation: the Gestapo leader of counterintelligence, Walter Schellenberg, was a prominent director and shareholder of ITT by arrangement with New York — an even Hitler dared not cross the Gestapo.[77]

The extent of ITT operations in Nazi Germany covered numerous sectors of the aviation industry, providing Navy spies ample opportunities to infiltrate German industry as Tompkins attests happened:

> ITT, through its subsidiary C. Lorenz AG, owned 25% of Focke-Wulf, the German aircraft manufacturer, builder of some of the most successful Luftwaffe fighter aircraft. In the 1960's, ITT Corporation won $27 million in compensation for damage inflicted on its share of the Focke-Wulf plant by Allied bombing during World War II. In addition, Sutton's book uncovers that ITT owned shares of *Signalbau AG, Dr. Erich F. Huth* (Signalbau Huth), which produced for the German Wehrmacht radar equipment and transceivers in Berlin, Hanover (later Telefunken factory) and other places.[78]

During World War II, Focke-Wulf, like many aviation companies, had to move its manufacturing plants underground to escape the Allied bombing campaigns. It was during this critical period that Focke-Wulf, an innovator in the aviation industry, became one of many German companies that moved assets and personnel to Antarctica. Focke-Wulf developed the first helicopters for the War effort and this made them one of the leading German companies for the construction of future spacecraft using advanced propulsion technologies.

Well after the end of WWII, Focke-Wulf rose as a prominent aviation company in Europe, and played a significant role in the formation of the European giant in Aerospace; Airbus. During this entire period, Focke-Wulf worked closely with ITT, which became an important U.S. collaborator in the development of a secret space program in Antarctica. Indeed, as will be discussed in chapter 14, ITT was among the first, if not THE first U.S. defense contractor to work in Antarctica. This is not all that surprising since according to Tompkins, ITT was a "Reptilian company", and Hitler had secretly reached agreements with the

Reptilians to move German resources to Antarctica, as will be discussed in detail in chapter four.[79] ITT was a key company in the worldwide Fraternity described by Higham, where loyalties transcended national allegiances to pursue a global elite agenda.[80]

In the lead up to World War II, while U.S. industries and bankers were flooding into Nazi Germany to sign merger deals, establish subsidiaries and build massive new factories, Hitler learned about the spacecraft prototypes that had been funded by the Thule and other German secret societies. The German Navy, through its close collaboration with Thule Society officials, was aware of its research and development of such promising spacecraft technologies bearing many similarities to submarine construction. The same slush fund, established by Wilhelm Canaris for secret submarine construction, was also available for such exotic research, which the German Navy was in the best position to secretly fund and develop. As mentioned in chapter 1, it was the German Navy that had arranged for scientists such as Professor Schumann to evaluate the free energy device developed by Hans Coler, and secretly develop working prototypes for submarine construction.

The same secret societies funding Hitler's rise to power now wanted him to provide state support for their efforts to build a secret space program capable of reaching the stars. Hitler was only too willing, and secretly committed the resources of the Nazi State into the research and development of such exotic technologies. His goals were quite simple. Technologies that could be developed to achieve flight to the stars could also be used to establish domination over planet Earth.

Leading German industrialists and Nazi ideologues agreed with the idea of developing the Vril spacecraft prototypes. The Nazi SS, led by Heinrich Himmler, would spearhead the effort to weaponize such advanced technologies for the coming war effort. At the same time, Hitler would provide all the technical and scientific assistance that was necessary for the German secret societies to build their interplanetary and interstellar space fleets.[81]

What is critical for understanding the development of spacecraft development programs in the Weimar Republic and in Hitler's Nazi Germany, is the role of leading German companies, the German Navy, and secret societies in their respective collaboration to fund and build these fleets. With the rise of the Nazi government, the Thule and other German secret societies would now have the unfettered support of the state in gaining the best scientific expertise and large scale funding for these top secret projects, along with the support of leading German companies. It is important to keep in mind that during this entire period, the close relationship between German and U.S. companies meant that "The Fraternity" was also very likely either a part of, or informed about, these events. And, it is very probable that U.S. companies, such as ITT, began collaborating with The Fraternity in developing spacecraft as early as the Weimar Republic years!

In early 1939, the German secret societies had decided that their space program would relocate to Antarctica, whose remoteness and deep cavern system under two miles of ice, accessible only by submarines, would provide all that was necessary to build such a program without possible disruption from the impending war. In the midst of these secret efforts, the role of U.S. companies must be considered as a significant factor in the development of two German space programs – the weaponization program in Nazi Germany, and the interplanetary/interstellar program in Antarctica. As already mentioned, the U.S. corporate involvement in Germany's burgeoning aerospace programs enabled an important back door for US Navy spies to penetrate these top secret operations, and report back to Naval Air Station, San Diego, about what was really happening in Nazi Germany and Antarctica.

CHAPTER 3

German Companies Begin Operations in Antarctica

German Bases Made Possible within the Inhospitable Continent

Extensive portions of the Antarctic were claimed by Captain Alfred Ritscher, on behalf of the Nazi government in 1938/1939, during the first Nazi expedition to Antarctica. While on this mission, the aircraft carrier, Schwabenland, sent planes to perform extensive aerial surveillance of the newly claimed region, which was named Neuschwabenland.[82] Among the goals of the Schwabenland expedition were the establishment of several bases in Antarctica. William Tompkins says US Navy spies-reported during their debriefings that the Nazis had been given extraterrestrial assistance guiding them to a large Antarctic cavern system, and specifically to where bases could be built in a favorable environment:

> Large portions of equipment were sent down there. But right next to them were three tremendous size caverns which the Reptilians had. Not Grays, but Reptilians. Germany got two more,

45

about a tenth the size of the big Reptilians [cavern]. They were able to ... [go] down, usually by submarine. They built these flat submarines, regular class, so they could ship all this stuff down.[83]

In a personal interview, Tompkins further adds:

The Reptilians already knew where all of the caverns were and all of the tunnels were. Okay? So this, again, is nothing new to the Reptilians, and giving instructions to the Germans, and it sort of comes up a little later as to where the good holes are underneath the water that go into tunnels because it's underneath the ocean's level. All of this was just handed over to them. They didn't have to do research.[84]

Corey Goode, who claims to have served in a "20 and back" US Navy secret space program, says he read a digital version of the briefing documents which Tompkins had prepared from the Navy spies' debriefings. Notably, this took place four decades later when Goode was searching the archives on an SSP smart glass pad. Goode remembers the documents specifically describing: "three known Antarctic (Cities/Bases) and several secret underground bases in Argentina" that had been established by the Nazis.[85]

Recently in 2017, Australian and New Zealand scientists *officially* discovered the existence of large natural caverns in Antarctica, thermally heated by adjacent volcanoes. It was found that the temperature in the caverns could be balmy 77° Fahrenheit (25°C) and support a variety of life forms! Chris Pash, a reporter with *Business Insider Australia*, wrote about the scientists' findings:

Around Mount Erebus, an active volcano on Ross Island in Antarctica, steam has hollowed out extensive cave systems. Dr. Ceridwen Fraser from the ANU [Australian National University] Fenner School of Environment and Society says forensic analyses of soil samples from these caves have revealed intriguing traces of DNA from algae, mosses and small animals. "It can be really warm inside the caves, up to 25 degrees Celsius in some caves," Fraser says.

... "The results from this study give us a tantalizing glimpse of what might live beneath the ice in Antarctica – there might even be new species of animals and plants," she says....

Dr. Charles Lee, another co-researcher from the University of Waikato, says there are many other volcanoes in Antarctica, so subglacial cave systems could be common across the icy continent.

"We don't yet know just how many cave systems exist around Antarctica's volcanoes, or how interconnected these subglacial environments might be. They're really difficult to identify, get to and explore," Dr. Lee says.[86]

The existence of extensive networks of mysterious underground caverns heated by nearby volcanoes is further strengthened by other recent scientific findings. In August 2017, scientists released a study revealing the discovery of 91 new volcanoes, in addition to the 41 previously found, thereby forming the world's most active volcanic region, deep under Antarctica's ice sheets. Robert McKie, a reporter with the *Guardian* newspaper wrote:

Scientists have uncovered the largest volcanic region on Earth – two kilometers below the surface of the vast ice sheet that covers west Antarctica.

The project, by Edinburgh University researchers, has revealed almost 100 volcanoes – with the highest as tall as the Eiger, which stands at almost 4,000 meters in Switzerland….

After the team had collated the results, it reported a staggering 91 previously unknown volcanoes, adding to the 47 others that had been discovered over the previous century of exploring the region.

These newly discovered volcanoes range in height from 100 to 3,850 meters. All are covered in ice, which sometimes lies in layers that are more than 4km thick in the region. These active peaks are concentrated in a region known as the west Antarctic rift system, which stretches 3,500km from Antarctica's Ross ice shelf to the Antarctic peninsula.

Geologists say this huge region is likely to dwarf that of east Africa's volcanic ridge, currently rated the densest concentration of volcanoes in the world.[87]

These fresh scientific findings support Tompkins and Goode's claims that the Germans had found or been led to a natural cavern system deep under the Antarctic ice that could provide a temperate environment to support a large base(s).

Goode also reports that he was taken on a secret tour of several of the natural caverns found under the Antarctic ice shelf in early 2016. He describes that one thermally heated cavern has been developed into a huge industrial area. Goode further

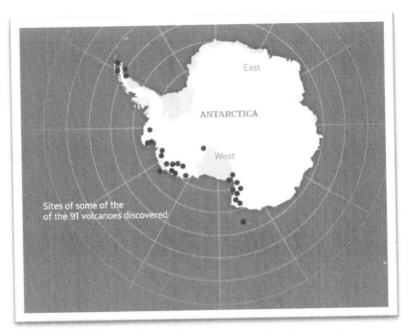

Figure 8. Newly discovered buried Volcanoes in Antarctica.

provides an artistic depiction of what he saw [see Figure 9]. More details of his alleged trip will be examined in chapter 11. Very significantly, the scientific findings just discussed were published **after** Goode made public his report of large city-sized caverns naturally heated by volcanic activity. Therefore, these recent findings provide solid independent scientific support for Goode's prior claims. All of the building and developments taking place under the ice is the direct result of the Germans being guided to one of more these extensive underground cavern systems prior to the start of World War II.

Back in the late 1930's, the establishment of Nazi Antarctic bases became a cause of real concern for President Roosevelt. He issued orders for the undertaking of a military expedition to challenge the Nazi's growing presence in regions of Antarctica regarded as part of the Western Hemisphere. *The New York Times* reported on July 7, 1939:

Figure 9. Illustration of Corey Goode's description of Secret Space Port under Antarctica. Courtesy of www.Gaia.com

President Roosevelt moved today to prevent possible extension of Germany's claims to Antarctic areas into the Western Hemisphere by directing Real Admiral Richard E. Byrd to leave in October to territory within the sphere of influence of the Monroe Doctrine ... it [is] apparent that this government was prepared to take the position, if necessary, that any attempts by foreign powers to establish bases west of the 180th meridian in the Antarctic would be considered an unfriendly act ...[88]

During World War II, extensive submarine activity in the region of Antarctica indicated that the Nazis were building more bases, in addition to the one Captain Ritscher's expedition had established. US Navy spies reported that once it became possible, major German companies involved in the Nazi's military industrial complex began moving equipment, resources and personnel down to Antarctica, using Nazi Germany's vast fleet of submarines, according to William Tompkins. Companies such as I.G. Farben, Krupp, Siemens, Messershmitt, United Steelworks, etc., were just some of many involved in the Antarctica operations which Tompkins specifically recalls from the Navy spies' intel.[89]

Having had access to this very same information from 1987 to 2007, Corey Goode is able to verify from memory reading similar things to points expressed in Tompkins' testimony. Goode confirms the German Secret Societies were in charge of the Antarctica operations, not Hitler's Nazi Government, with leading corporations involved in the construction activities taking place there – using forced labor.[90]

The vast Antarctica operations were orchestrated by Admiral Wilhelm Canaris, head of German Military Intelligence (1935-1944), who used his previous experience in the covert rebuilding of Germany's submarine fleets after World War I to do something very similar in Antarctica. This time it would be spacecraft, and not just submarines, that Canaris would arrange to be built using an international network of German companies and banks, along with their foreign partners. As mentioned earlier, Canaris worked closely with the Thule Society in achieving its long term goals, with debate only remaining as to whether or not he was a full member.

Rudolf Hess, and other well-known Thule Society colleagues who had infiltrated the Nazi Party, worked closely with Canaris in directing substantial resources down to Antarctica. The German Navy was a key component at this time, since only it could provide the large cargo submarines Type X (XB) U-boats which were built by a Krupp subsidiary in Kiel) to move all that was necessary down to Antarctica. More importantly, the inherent similarities between submarine and spacecraft construction, plus the congruent crew training for both types of vehicle operations, made the German Navy the military service most suited to lead Antarctic operations.

The goals of the Thule Society/Nazis/German Navy were

threefold when it came to Antarctica. The first was to locate and convert vast caverns found deep under the Antarctic ice into fully equipped bases that could host large numbers of personnel and equipment. Secondly, to establish large manufacturing plants in Antarctica for building fleets of spacecraft capable of interplanetary and, eventually, interstellar flight. Finally, the last goal was to establish Antarctica as a safe haven from the vicissitudes of another European war.

Hitler was carefully steered into believing that the impervious bastion being built in Antarctica would ensure his personal future welfare. Certainly, this is what Admiral Karl Donitz implied when he bragged about the remote (Antarctica) base on three separate occasions, and how it would provide a safe haven for Hitler if he ever required it.

In 1943, Donitz is reported to have made his first statement: "the German submarine fleet is proud of having built for the Führer, in another part of the world, a Shangri-La on land, an impregnable fortress."[91] The second occasion was in 1944, when he revealed how plans were in place to relocate Hitler so he could launch a new effort for his thousand-year Reich:

> The German Navy will have to accomplish a great task in the future. The German Navy knows all hiding places in the oceans and therefore it will be very easy to bring the Führer to a safe place should the necessity arise and in which he will have the opportunity to work out his final plans.[92]

Donitz's reference to the German Navy's knowledge of "all hiding places in the oceans" shows once again how central the Navy was to the entire Antarctica operation.

Finally, Dönitz's remarks at his Nuremberg war crime trial clearly identified Antarctica as the place where Germany's most advanced technologies had been secretly relocated by their Navy's large submarine fleet. At the trial he boasted of "an invulnerable fortress, a paradise-like oasis in the middle of eternal ice."[93]

A world famous Austrian cartographer and artist, Professor Heinrich C. Berann, provides exciting proof that Donitz's statements are entirely plausible. Berann worked for the National Geographic Society beginning in 1966, and later Colombia University and the US Navy, creating ocean floor maps. In 1972, he rendered a map of Antarctic without its mantle of ice, notably displaying underwater passageways that ran throughout the continent.[94] This distinguished map substantiates the naturally existing routes in which submarines could travel under the ice for considerable distances, to Nazi Germany's "invulnerable fortress", via a natural cavern system buried under nearly two miles of ice in places.

A more recent detailed map of what lies under the ice sheets of Antarctica was provided by the U.S. National Science Foundation in 2013.[95] Extensive river systems and lakes were found under the ice sheets, once again affirming a plausible means of navigating under Antarctica, just as Tompkins and Goode have described (see Figure 11).

Admiral Donitz's claims are further supported by documents, provided by an alleged German submarine crewman after the war, which give detailed instructions for U-Boat Captains to reach the Antarctica bases through the hidden passageways. Figure 12 displays an image of the document with the translated instructions.[96]

In the supply and secret construction of the Antarctica bases through the Kriegsmarine's submarine fleet, Admiral Donitz worked closely with Admiral Canaris. When Hitler relieved Canaris of his post as chief of the German's military intelligence service (the Abwehr) and transferred its control to Himmler's SS in March 1944, Donitz reassigned Canaris to exclusive service with the navy.[97] Canaris would now concentrate on supporting the wholescale movement of resources and personnel for the secret

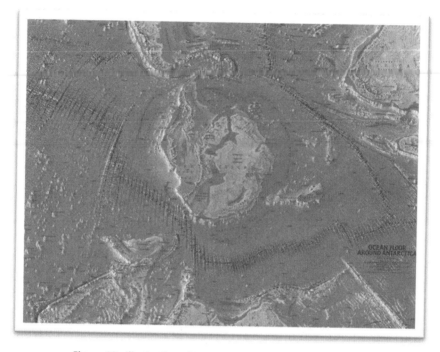

Figure 10. Illustration of Ice Free Antarctica by Heinrich Berann

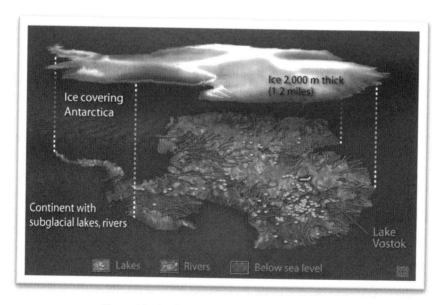

Figure 11. Credit: U.S. National Science Foundation

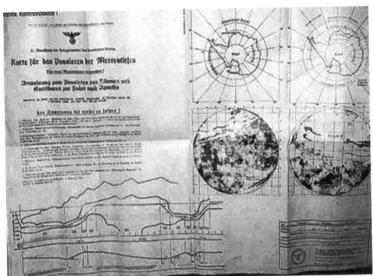

ecent at the point with the coordinates: Exact Intersection: 64° southern latitude and 1° eastern ongitude, to a depth of 400 meter.

he instruction have to be strictly followed!

1. Decent, from the point of decent with half speed, a starboard declination of 10° with a bow-heaviness declination angle of 5°. Distance 188 sm. Given depth – 500 meter. (Because of the moving inside the corridor the pressure on the ship body when maneuvering is insignificant)

2. Ascent. Full load with a stern trim. Ascent angle 23° with a port declination of 22°. 190 meter upwards. Distance 75.5 sm.

3. Difficult maneuver! Ascent full speed with a stern trim. Ascent angle 41°. Proceed straight ahaead. 110 meter upwards, distance 21.5 sm. Afterwards starboard declination of 8° until ascent to the surface in a distance of 81sm.

4. Proceed on the surface within the grotto with a starboard declination of 8°, Distance 286 sm.

5. 6. Schwieriges maneuver! Descent. With a bow-heaviness declination 45° to a depth of 240 meter. Distance 60 sm. Afterwards with a port declination of 20°, at which the descent to 310 meter to the entrance to the corridor continues. After the 310 meter mark the bow heavy descent need to be continued. Descent angle 7° until 360 meter, distance 70 sm. Futher starboard declination of 31° to a depth of 380 meter.

6. Descent. Bow heavy, ascent angle 22°, 100 meter upwards with a port declination of 26°. Distnace 43 sm.

7. Ascent. Stern trim, Ascent angle 45°, straight ahead until reaching the surface of Agartha. Distance 70 sm.

8. Proceed to Agartha. Full Speed. Proceed straight ahead, until the new light can be seen. Change of magnetic poles. The changes of the compass needle and instruments are to be disregarded.

(Further instructions in package Nr. 3 only when arrived in Agartha to be opened)

Figure 12. Directions to Antarctic bases – translated from original in German.

Antarctica facilities. On July 1, 1944, Canaris was appointed by Hitler to head the "Special Staff for Anti-Shipping Warfare and Economic War Measures", where he concentrated on *Operation Eagle Flight*; the transfer of industrial assets and Nazi capital to neutral locations and Antarctica.[98]

According to Michael Mueller, author of an authoritative biography on Canaris, after the failed assassination attempt against Hitler on July 20, 1944, Canaris was implicated as the "spiritual founder of the resistance movement."[99] He was arrested three days later, and after his personal diaries were found in the beginning of April 1945, he was summarily tried and then executed at Flossenberg Concentration Camp on April 7.[100]

William Tompkins, however, contradicts this scenario and says the US Navy spies reported this was all a sham – Canaris was secretly taken to Antarctica to play a leading role in directing future Antarctica operations. If Canaris' imprisonment, trial and execution were indeed staged, then it was a brilliant ploy in misdirecting Allied attention away from his critical role in setting up, supplying, funding and later leading the Antarctica bases established by German Secret Societies.

German Companies Begin Flying Saucer Production

Simultaneous flying saucer programs had been developed by Nazi Germany: one in Occupied Europe and the other in Antarctica. These programs were coordinated during the development and construction process for the creation of the 30 prototype antigravity spacecraft the Navy spies came to report about in their debriefings, as Tompkins explains:

> They built the prototypes in Germany. They built pre-protype, something which is ready for production, in Antarctica. They put this stuff in

production in the countries all over Germany [Occupied Europe], and they continued to build similar vehicles in Antarctica.[101]

There have been a number of authors that have analyzed the flying saucer reports from Nazi Germany, and the development of these craft in classified facilities. Henry Stevens' book, *Hitler's Flying Saucers* (2003), provides the most comprehensive overview of the numerous newspaper reports and official documents that have appeared or surfaced.[102] Stevens and other authors have also included key interviews with scientists such as Giuseppi Belluzo and Rudolph Schriever, who each almost simultaneously went on the public record in March 1950 about their participation in German flying saucer programs.[103]

In addition, Stevens has analyzed "smoking gun" FBI documents containing interviews with credible witnesses of such craft, helping to prove that the Germans were indeed building and testing saucer craft.[104] Among these smoking gun documents are several involving a Polish immigrant living in Texas, who describes his wartime experience of seeing a German flying saucer in a secure enclosure in 1944, while he was being held as a POW in Germany. A November 7, 1957 FBI Teletype gave a summary of the interview:

UPON INTERVIEW ADVISED THAT WHILE GERMAN POW DURING NINETEEN FORTYFOUR OBSERVED A VEHICLE DESCRIBED AS CIRCULAR IN SHAPE, SEVENTY FIVE TO ONE HUNDRED YARDS IN DIAMETER, APPROXIMATELY FOURTEEN FEET HIGH. THE VEHICLE WAS OBSERVED TO SLOWLY RISE VERTICALLY TO HEIGHT SUFFICIENT TO CLEAR FIFTY FOOT WALL AND TO MOVE SLOWLY HORTIZONTALLY A SHORT DISTANCE OUT OF VIEW [105]
...

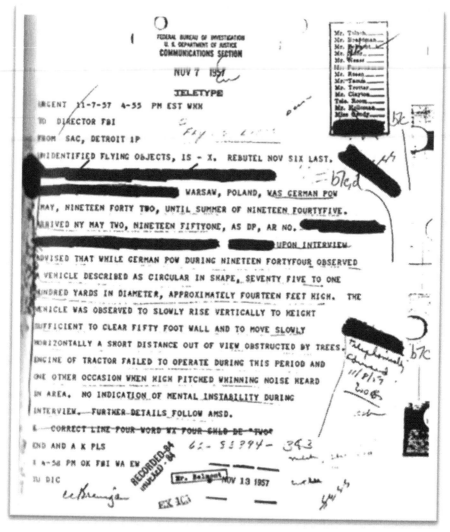

Figure 13. FBI Teletype discussing Polish witness of German Flying Saucer

Additionally, the CIA took a keen interest in foreign newspaper reports of flying saucer sightings or, more specifically, Nazi Germany's development of such craft. One CIA document, dated 12 January 1954, gives a summary of famed German engineer, George Klein's overview of the different flying saucer projects he worked on in Nazi Germany:

A German newspaper [not further identified] recently published an interview with George Klein, famous German engineer and aircraft expert, describing the experimental construction of "flying saucers" carried out by him from 1941 to 1945. Klein stated that he was present when, in 1945, the first piloted "flying saucer" took off and reached a speed of 1,300 miles per hour within 3 minutes. The experiments resulted in three designs: one, designed by Miethe, was a disk-shaped aircraft, 135 feet in diameter, which did not rotate, another designed by Habermohl and Schriever, consisted of a large rotating ring, in the center of which was a round, stationary cabin for the crew. [106]

What these FBI documents and newspaper reports clearly demonstrate is that the Germans were developing multiple flying saucer craft as part of the war effort. These sources give much information about the development and testing of prototypes, but little information about the successfulness of any program. In fact, the overall conclusion emerging from official documents and newspaper reports is that the Germans had failed in developing any flying saucer prototypes that could be successfully operated, let alone used for the war effort.

Information about the overall scope of German saucer programs, and the companies involved in any successful production models, has proven elusive. This is due to intelligence agencies from both NATO and former Warsaw Pact nations keeping a tight hold on official documentation that would show any success by the Germans. That situation changed dramatically with a sequence of events that led to the official end of the Warsaw Pact on February 25, 1991. Intelligence files were leaked and sold to the highest bidder by former intelligence operatives

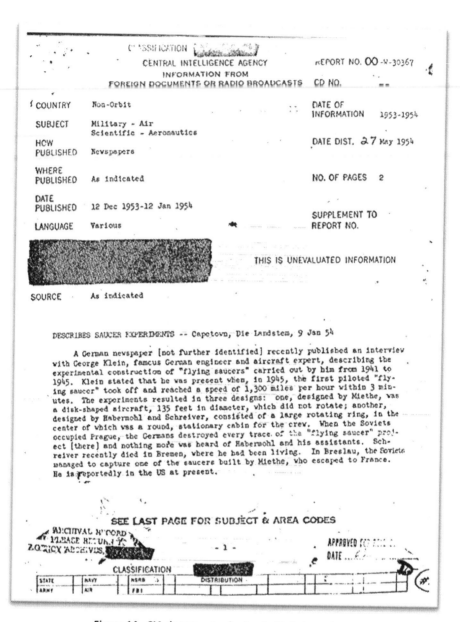

Figure 14. CIA document referring to Klein interview

now scrambling to financially survive in the chaos emerging from the collapse of the communist system. One of the former Warsaw Pact member states was Bulgaria, and members of its Academy of

Sciences would certainly have been among those approached by intelligence operatives eager to dispose any science related files in their possession.

Vladimir Terziski, a trained engineer and physicist, was a former member of the Bulgarian Academy of Sciences before emigrating to the U.S. in 1984.[107] He says that in 1991, he came into the possession of a leaked documentary film from the Nazi SS archives revealing different types of flying saucer craft built in Germany.[108] Based on Terziski's film, which he showed at public seminars from 1992 onwards, a number of documents began circulating describing the testing and manufacturing of Nazi Germany's successful flying saucer prototypes.

One document describes the number of different Haunebu and Vril craft built for the war effort, the number of test flights, and the propulsion systems then available (see Fig 15). All the craft were assembled at a remote German location designated as "Hauneburg", which was later shortened to Haunebu from which the successful prototypes got their designations. According to Nazi UFO historical researcher Rob Arndt, the Huaneburg site was chosen in 1935 by the Thule Society, but by 1942 the location was abandoned due to the changing war situation.[109] Since the "Vril 1" was the first flying saucer craft to be developed in Nazi Germany, it's worth beginning this overview of the 17 craft mentioned in the document, which were flight tested over 80 times.

In his lectures, Terziski discusses the testing and operational performance of the Vril 1 craft, according to details contained in one of the Nazi SS documents:

> The first purely Vril disc — the Vril 1 Jager (Hunter) was constructed in 1941 and first flew in 1942. It was 11.5 meters in diameter, had a single pilot, and could achieve 2,900 km/h – 12,000 km/hr. It flew with a metal dome at first but subsequent test

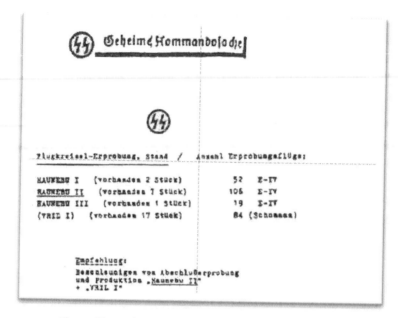

Figure 15. Production statistics for German Flying Saucers

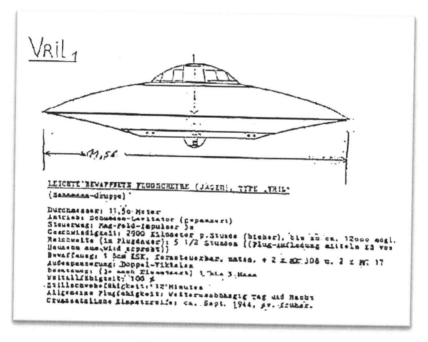

Figure 16. Vril 1 Specifications

versions had a heavily reinforced glass dome and could seat two crew. Flight endurance was 5.5 hrs. It was planned to arm this craft with two MK-108 cannon plus 2 MG-17 machineguns. Seventeen of these craft were constructed and tested between 1942-44 with 84 test flights.[110]

Next, the Nazi SS document states that two "Haunebu I" flying saucer craft were built and test flown 52 times. According to Terziski's information, the Haunebu I began testing in 1939, which Rob Arndt explains further:

> The early Haunebu I craft of which two prototypes were constructed were 25 meters in diameter, had a crew of eight and could achieve the incredible initial velocity of 4,800 km/h, but at low altitude. Further enhancement enabled the machine to reach 17,000 km/hr. Flight endurance was 18 hours. To resist the incredible temperatures of these velocities a special armor called Victalen was pioneered by SS metallurgists specifically for both the Haunebu and Vril series of disc craft. The Haunebu I had a double hull of Victalen. The early models also attempted to test out a rather large experimental gun installation – the twin 60mm KraftStrahlKanone (KSK) [a cannon] which operated off the Triebwerk [engine] for power. It has been suggested that the ray from this weapon made it a laser, but it was not.[111]

I have yet to find any information on a metal alloy called "Victalen", which only appears in sources that can be traced back to Terziski. However, Henry Stevens investigated claims about advanced metal alloys being produced in Germany's advanced aviation projects. He found substantial evidence that esoteric metal alloys such as "Impervium" and "Lubricium" were indeed being produced, and some of these were brought to the U.S. for further development in classified facilities.[112] Stevens cites a metallurgist from the former aerospace company TRW (now part of Northrup Grumann) who responded to a question about whether such "super metals" had been created by Nazi Germany:

> It's true. The Germans developed all sorts of alloys during the war. After the war we took them – some of them were great – we took one and gave it a TRW number, and still market it today – we didn't want to give the Germans credit though.[113]

It's also worth recalling that Tompkins, who worked at TRW from 1967 to 1971, says that the Germans were helped by Reptilian extraterrestrials in advanced aerospace projects, including metallurgy. Consequently, major German steel companies such as Thyssens AG and Krupp were very likely heavily involved in developing unique metal alloys for the flying saucer programs. If so, it remains a trade secret or classified to the present day, as the TRW metallurgist expressed. Importantly, in 1999 Thyssens and Krupp merged to form one of the world's largest steel producers. ThyssenKrupp AG almost certainly continues to play a key production role in the secret space program started by the Germans in Antarctica.

It's worth contrasting the top speed of the successful Haunebu I craft (4,800 to 17,000 km/hr) with the model George Klein witnessed being tested in 1945 (1,300 mph/2,200 km/hr). Clearly, there is a vast performance difference between the successful and unsuccessful flying saucer prototypes under research and development.

According to Terziski's Nazi SS saucer production document, seven Haunebu II were built and test flown 106 times. He provided additional details on its performance, as Arndt illustrates:

> In 1942, the enlarged Haunebu II of 26 meters diameter was ready for flight testing. This disc had a crew of nine and could also achieve supersonic flight of between 6,000-21,000 km/h with a flight endurance of 55 hours. Both it and the further developed 32 meter diameter Do-Stra [Dornier Stratosphärenflugzeug] had heat shielding of two hulls of Victalen. Seven of these craft were constructed and tested between 1943-44. The craft made 106 test flights. [114]

Significantly, Terziski has identified Dornier Flugzeugwerke (Aircraft Plants) as the German corporation responsible for the successful Haunebu II prototype.

> By 1944, the perfected war model, the Haunebu II Do-Stra [Dornier Dornier Stratosphärenflugzeug] was tested. Two prototypes were built. These massive machines, several stories tall, were crewed by 20 men. They were also capable of hypersonic speed beyond 21,000 km/h. The SS had intended to produce the machines with tenders for both Junkers and Dornier but in late 1944/early 1945 Dornier was chosen. [115]

Dornier operated as a privately held company from 1914 to 1996, when it was acquired by the Fairchild Aircraft company in 1996, and eventually absorbed into Airbus. [116] Terziski's document reveals that Dornier's successful model was put into production.

Next, we have the Haunebu III which Terziski's document asserts only one prototype was built and tested 19 times. Again, Arndt summarizes Terziski's information:

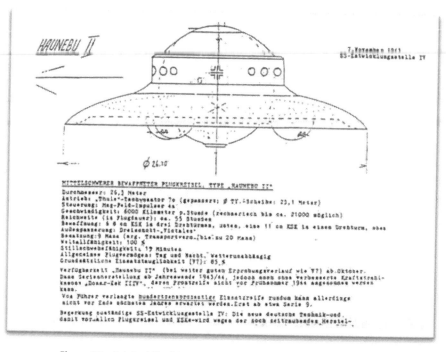

Figure 17. Original Nazi SS Document of Haunebu II specifications

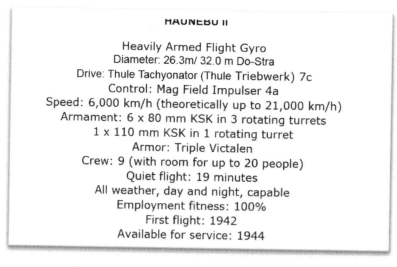

HAUNEBU II

Heavily Armed Flight Gyro
Diameter: 26.3m/ 32.0 m Do-Stra
Drive: Thule Tachyonator (Thule Triebwerk) 7c
Control: Mag Field Impulser 4a
Speed: 6,000 km/h (theoretically up to 21,000 km/h)
Armament: 6 x 80 mm KSK in 3 rotating turrets
1 x 110 mm KSK in 1 rotating turret
Armor: Triple Victalen
Crew: 9 (with room for up to 20 people)
Quiet flight: 19 minutes
All weather, day and night, capable
Employment fitness: 100%
First flight: 1942
Available for service: 1944

Figure 18. Translation of Hanuebu 11 Specifications

Yet larger still was the 71 meter diameter Haunebu III. A lone prototype was constructed before the close of the war. It was crewed by 32 and could achieve speeds of between 7,000 - 40,000 km/hr. It had a triple Victalen hull. It is said to have had a flight endurance of between 7-8 weeks! The craft made 19 test flights. This craft was to be used for evacuation work for Thule and Vril in March 1945.
[117]

It's important to note that the Earth's escape velocity is 40,270 km/hr (25,020 mph), which means that the Haunebu III was capable of leaving Earth's orbit. It is therefore the world's first spacecraft.

Terziski explains that Vril and Haunebu models possessed an electro-gravitics propulsion system called Thule-Tachyonator drives. These were first developed in 1939 by the Nazi SS E-IV development unit, which was part of the "Order of the Black Sun" according to researcher Rob Arndt:

> This group developed by 1939 a revolutionary electro-magnetic-gravitic engine which improved Hans Coler's free energy machine into an energy Konverter coupled to a Van De Graaf band generator and Marconi vortex dynamo [a spherical tank of mercury] to create powerful rotating electromagnetic fields that affected gravity and reduced mass. It was designated the Thule Triebwerk [Thrustwork, a.ka. Tachyonator-7 drive] and was to be installed into a Thule designed disc.[118]

Terziski contends that by 1943 the German companies Siemens and AEG had built assembly lines for mass production of the Thule-Tachyonator drives that powered the Vril and Haunebu flying saucers, which were soon moved to Antarctica.

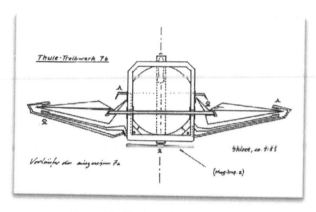

Figure 19. Thule-Tachyonator Drive

Siemens played a critical role in building the Hans Coler free energy device that powered not only the German Navy's most advanced submarines, but its secret fleet of flying saucers. Henry Stevens, author of *Hitler's Suppressed and Still-Secret Weapons, Science and Technology*, writes:

> The Magnestromapparat was fully developed and completed by 1933 with the help of von Unuh and Fraz Haid of Siemens-Schukert. This means that the Siemens firm, in spite of later denials (or memory losses) has known of free energy and specifically the Hans Coler device since the early 1930s.... In 1943 Coler and his work, ended up at the O.K.M or the German Navy.[119]

Finally, when it comes to advanced weapons systems, energy production and space medicine, the industrial conglomerate I.G. Farben played the vital role in overseeing these areas for the two German space programs. I.G. Farben has been a pioneer in many advanced research topics, including being among the first companies to develop laser weapons, which were supplied for the successful flying saucer prototypes that were eventually sent down to Antarctica. Henry Stevens uncovered documents showing that the Allied powers had learned about I.G. Farben scientists being involved in the development and testing of laser weapons that produced devastating effects:

> It is also interesting to note that the Allies were given a list of scientists who participated in the laser experimentation at I.G. Farben ... Besides the detail of the layout, it is most impressive to note that when the test targets of the weapon, rats, were hit with the ray, their bodies glowed for a fraction of a second before they completely disintegrated. [120]

These laser weapons were used in their ruinous capacity (as will be shown in the next chapter) against the 1946/1947 Operation Highjump mission launched by the US Navy.

In addition, I.G. Farben was internationally known for its pioneering medical research, which was essential for enabling German astronauts to safely fly into space.[121] Significantly, Tompkins claims that the Reptilians supplied the Germans with information on advanced medical subjects such as cloning and age regression, which the Nazis invested significant resources into developing for the war effort.[122] I.G. Farben's Nobel Prize winning scientists would have been among those tasked to initiate the research and development of these and other esoteric medical subjects. Therefore, as the war effort faltered, I.G. Farben was among the companies instructed to take resources out of Germany and its scientists were included with those secretly moved to Antarctica.

It was the Thule Society, and their secret society brethren, who had thoroughly infiltrated the Nazi government and were actually running the Antarctica operations, while also maintaining control of the German companies that built the prototype antigravity spacecraft. The key Nazi official who oversaw the entire Antarctica operation from its inception in 1938/1939 was Admiral Canaris. His principal loyalty was never to Hitler, however, but to the nationalist agenda espoused by German Naval Intelligence and the Thule Society, which increasingly came into conflict with Hitler's reckless militaristic policies that culminated in World War II.

This is why leading industrialists such as Fritz Thyssen opposed Hitler's decision to invade Poland and start World War II, and why Canaris led efforts behind the scenes to depose Hitler during the 1938 Czech crisis. While the Thule Society and German Naval Intelligence supported Hitler's rise to power and could influence his policies in a nationalist direction, they could not fully control him when it came to larger issues of war and peace. Their best option in this difficult situation was to independently develop the Antarctica operations, and exclude Hitler and his Nazi SS henchmen from controlling those operations.

As the tide of World War II turned, Hitler became desperate and demanded that the "wonder weapons" which were being developed in Antarctica be used for the War effort. According to Corey Goode, the German Secret Societies refused:

> I think that there were [some] technologies acquired and integrated into their breakaway secret space program that they were developing. But they were developing this for their own [purposes] when it came down to it, they didn't care about [Germany winning] World War II, the motherland, [or] using this technology to defeat the United States and the enemies they were engaged in war with.[123]

When Hitler learned, through his Nazi SS subordinates, that the Antarctica operations had gone rogue and would not provide advanced weapons systems for the war effort, he became furious. In February 1944, Canaris was removed from power, and the

German Abwehr (military intelligence) was soon after absorbed into Himmler's SS.

However, The Thule Society succeeded in establishing Antarctica as a powerful fortified base, outside of Hitler's control because of their able handlers, who ensured that he would not interfere with their plans there for building antigravity spacecraft capable of interplanetary and interstellar operations. The first handler of Hitler up to 1941, was his deputy Führer, Rudolf Hess. Hess' career came to an end, as previously discussed, after he attempted to end the war with the United Kingdom through secret negotiations with the British aristocracy, but was captured when the Churchill faction learned of the peace mission.[124]

Hitler's second handler was Hess' replacement, Martin Bormann, who assumed all the functions of Deputy Fuehrer, which placed him in the role of chief of the newly created office of Party Chancellery. Bormann also worked closely with the Thule Society and German industrialists who well understood how Hitler's extremism was leading Germany to military and financial ruin. Bormann was the key Nazi official who assisted German industrialists in an elaborate "capital flight plan", assuring that a Fourth Reich would emerge out of the ashes of World War II.

Financial Foundations for a Fourth Reich and Antarctic Development

After early military successes, the Nazis began experiencing major setbacks in the war against the Soviet Union. Their defeat at Stalingrad in February 1943, during which the entire German 6th Army was destroyed, signified that the tide had decisively turned and the Soviet Union was going to prevail in the war. German industrialists saw the writing on the wall and made preparations for the movement of large financial resources to safe locations in various neutral countries, South America and

Antarctica.

Curt Reis, a well-known news correspondent during World War II, describes the first meeting held by industrialists who began to prepare for the likely defeat.

> In May 1943, in the wake of the defeat at Stalingrad, Reiss said German industrialists met in Chateau Huegel near Essen, home of the Krupps, and reviewed the situation of their nation. The decision was to distance German commerce from the Nazi regime, Reiss wrote, adding: "All future changes discussed at the meeting centered around the idea of divorcing German industry as far as possible from Nazism as such. Krupp [von Bohlen und Halbach] and [I. G. Farben Director Georg von] Schnitzler declared that it would be much easier for them to work after the war if the world were certain that German industry was not owned and run by the Nazis. He said that Goering as well as other influential party men saw eye to eye with him on this, and would consent to any arrangement that did not involve the prestige of the party."[125]

Admiral Canaris, with the help of Bormann, assisted the worried German industrialists in their capital flight plans. Canaris was an old hand in moving large amounts of capital around the world, and Bormann quickly learned what needed to be done.

After the Allied Powers successfully landed in Normandy on June 6, 1944, and established a long anticipated beachhead for the Western Front, Bormann and leading industrialists became desperate to quickly move capital and resources out of Germany to neutral international locations and Antarctica. On August 10, 1944, Bormann secretly convened a meeting of leading German industrialists and instructed his emissary, SS Obergruppenfuhrer (equivalent to Lieutenant General) Dr. Scheid, to tell them the war was lost.

A US Army Intelligence File, called the "Red House Report", provides important details into the German companies that worked with Bormann in the Nazi capital flight plan. This document, dated November 7, 1944, describes how German industrialists were told to evacuate all available assets to neutral countries using hundreds of shell companies designed to hide the massive out-flow of Nazi capital and industrial resources.[126]

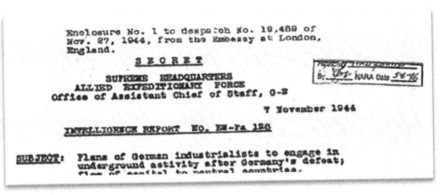

Figure 20. Red House Report

The ultimate source of secret orders approving the Nazi capital flight, according to Paul Manning, author of the book *Martin Bormann: Nazi in Exile,* was Bormann; whose influence over the Nazi Party increasingly grew as Hitler became despondent over an impending military defeat.[127] This fact is exemplified in Scheid's involvement as SS, which reveals that Himmler was also aware that Bormann had the reins of authority to launch such an initiative, and that the Nazi SS would support Bormann.

Bormann was explicit in his instructions of how financial assets must be sent out of Germany, and this plan was relayed through Scheid:

> From now on also German industry must realize
> that the war cannot be won and that it must take
> steps in preparation for a post-war commercial
> campaign. Each industrialist must make contacts
> and alliances with foreign firms, but this must be
> done individually and without attracting any
> suspicion. Moreover, the ground would have to be
> laid on the financial level for borrowing
> considerable sums from foreign countries after the
> war. [128]

The greater goal of Bormann's plan was to enable and ensure by covert economic means the emergence of a new German Empire, a Fourth Reich, as also described in the report:

> [I]t was stated that the Nazi Party had informed the
> industrialists that the war was practically lost but
> that it would continue until a guarantee of the
> unity of Germany could be obtained. German
> industrialists must, it was said, through their
> exports increase the strength of Germany. They
> must also prepare themselves to finance the Nazi
> Party which would be forced to go underground as
> Maquis (in Gebirgaverteidigungastellen gehen).
> From now on the government would allocate large
> sums to industrialists so that each could establish a
> secure post-war foundation in foreign countries.
> Existing financial reserves in foreign countries must
> be placed at the disposal of the Party so that a
> strong German Empire can be created after the
> defeat. [129]

Bormann arranged for the creation of 750 front companies for the Nazi capital flight plan called *Operation Eagle*. Both Fritz

Thyssen and the I.G. Farben conglomerate were vital to this operation:

> As part of this plan, Bormann, aided by the black-clad SS, the central Deutsche Bank, the steel empire of Fritz Thyssen, and the powerful I. G. Farben combine, created 750 foreign front corporations – 58 in Portugal, 112 in Spain, 233 in Sweden, 214 in Switzerland, 35 in Turkey, and 98 in Argentina. [130]

The giant chemical conglomerate I.G. Farben, which in itself formed a state within a state, had extensive international partners that were used for the capital flight plan:

> [U.S.] Treasury investigations discovered Farben documents that showed the firm maintained an interest in more than 700 companies around the world. This number did not include Farben's normal corporate structure, which covered ninety-three countries, nor the 750 corporations created under Bormann's flight capital program. I. G. Farben also was at the hub of money transfers out of Nazi Germany. Even before the end of the war, for example, "I. G. Latin American firms all maintained, unrecorded, in their books, secret cash accounts in banks in the names of their top officials," wrote Manning. These were used to receive and to disburse confidential payments; firms dealing with Farben wanted this business but certainly did not wish it known to British and United States economic authorities." [131]

A US Army Report prepared for the prosecution of I.G. Farben executives at the Nurenburg War Crimes describe how

Farben had become adept at hiding its assets internationally, and exercising control through a variety of economic tools:

> The company cloaked its direct and indirect ownership and control of hundreds of its foreign subsidiaries by utilizing every conceivable device known to the legal and "extra legal" mind, including the use of nominees, option agreements, fictions or intervening transfers, dividend and loan agreements, pool agreements, endorsements in blank escrow deposits, pledges, collateral loans, rights of first refusal, management contracts, service contracts, patent agreements, cartels and withholding know-how. Geheimrat Hermann Schmitz, I.G's president, was known throughout the industrial world as "the master of financial camouflage."[132]

It's worth mentioning that the Dulles' brothers, through their legal firm, Sullivan and Cromwell, helped I.G. Farben establish its shady international financial network prior to WWII.

Fritz Thyssen, though under Nazi House Arrest due to his opposition to Hitler's war policies, was heavily involved in *Operation Eagle Flight* (Aktion Adlerflug).[133] His old U.S. partners and associates were invaluable in helping Nazi capital reach safe locations:

> Bormann's Operation Eagle Flight was substantially helped by the close connections with foreign banks and businesses begun long before the war. According to former U.S. Department of Justice Nazi War Crimes prosecutor John Loftus, much of the wealth was passed out of Germany by German banker Fritz Thyssen through his bank in Holland, which, in turn, owned the Union Banking

Corporation (UBC) in New York City.[134]

Thyssen used his Dutch bank to transfer Nazi capital, with help from his New York associates, which included Prescott Bush:

> Thyssen did not need any foreign bank accounts because his family secretly owned an entire chain of banks. He did not have to transfer his Nazi assets at the end of World War II, all he had to do was transfer the ownership documents – stocks, bonds, deeds, and trusts – from his bank in Berlin through his bank in Holland to his American friends in New York City: Prescott Bush and Herbert Walker. Thyssen's partners in crime were the father and father-in-law of a future president of the United States."[135]

The success of the Nazi capital flight program is revealed by Paul Manning's description of its impact on the modern era:

> [T]he 750 new corporations established under the Bormann program, gave themselves absolute control over a postwar economic network of viable, prosperous companies that stretched from the Ruhr to the "neutrals" of Europe and to the countries of South America; a control that continues today and is easily maintained through the bearer bonds or shares issued by these corporations to cloak real ownership.[136]

According to Manning, Allied governments turned a blind eye to Nazi capital flight since there were significant financial benefits for major U.S. and British corporations:

> They had understandable reasons, if you overlook

morality: the financial benefits for cooperation (collaboration had become an old-hat term with the war winding down) were very enticing, depending on one's importance and ability to be of service to the organization and the 750 corporations they were secretly manipulating, to say nothing of the known multinationals such as I.G. Farben, Thyssen A.G., and Siemens.[137]

Another key factor to keep in mind here is that Allen Dulles, who ran the Swiss headquarters of the Office of Strategic Services (forerunner to the CIA) was orchestrating secret deals with Nazi authorities that would allow significant war resources to fall into U.S., rather than Soviet, hands. Indeed, the capture of U-boat 234 carrying enriched uranium was likely part of a deal where the U.S. would be assisted in developing atomic bombs, in exchange for not interfering with Operation Eagle Flight, and allowing prominent Nazis such as Hitler and Bormann to escape.[138]

Martin Bormann was able to escape to Argentina after the end of hostilities in Europe in May 1945, as documented by multiple researchers, including historian Paul Manning.[139] In addition to Bormann, Adolf Hitler also escaped to Argentina and lived out his exile years in the relative comfort of Bariloche, according to a number of official government documents and eyewitness testimonies assembled by historians such as Harry Cooper.[140] In October 2017, newly declassified CIA files contained a new batch of Kennedy Assassination documents, which remarkably included reports of Hitler visiting Colombia and Argentina!

In one document, a former Nazi SS Stormtrooper, Philip Citroen, reports having met with Hitler while visiting Columbia in 1955. He discusses a photograph taken of Hitler while in Columbia, who at the time was using a pseudonym. It is the only known photo of Hitler taken after the end of World War II that shows he both survived and was living in exile in South America.

In the document, a CIA source with the codename CIMELODY-3 reports on what she/he learned from a trusted friend:

> CIMELODY-3's friend stated that during the latter part of September 1955, a Phillip CITROEN, former German SS trooper, stated to him confidentially that Adolph HITLER is still alive. CITROEN claimed to have contacted HITLER about once a month in Colombia on his trip from Maracaibo to that country as an employee of the KNSM (Royal Dutch) Shipping Co. in Maracaibo. CITROEN indicated to CIMELODY-3's friend that he took a picture with HITLER not too long ago, but did not show the photograph. He also stated that HITLER left Colombia for Argentina around January 1955. CITROEN commented that in as much as ten years have passed since the end of World War II, the Allies could no longer prosecute HITLER as a criminal of war.[141]

It's important to keep in mind that the CIA has withheld the files from public release for over 50 years since President Kennedy's assassination on national security grounds. It is plausible that their release was considered dangerous as it raised a possible connection between a rumored 4th Reich and the Kennedy Assassination, which will be explored further in chapter seven.

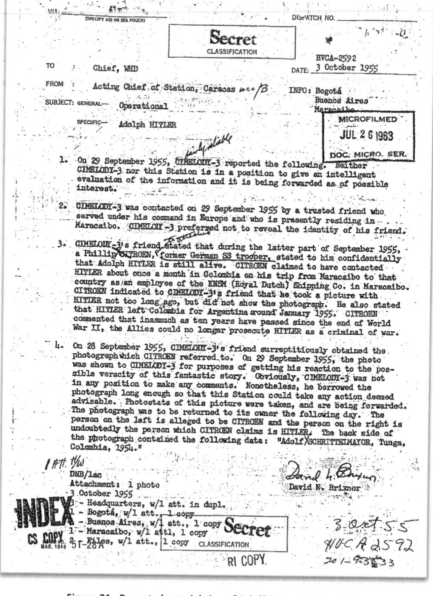

Figure 21. Report about sighting of Adolf Hitler in Colombia

Figure 22. Alleged photo of Adolf Hitler in Colombia in 1955

CHAPTER 4

The Nazi Retreat to Antarctica & South America

The Antarctic Exodus

The 'official' defeat of Nazi Germany in May 1945 was in fact only a 'tactical victory' masking a major strategic defeat withstood by the 'victorious Allies'. This potentially devastating secret was kept from the general public in the decades to follow. The Allies would not admit that a significant proportion of Nazi Germany's political elite, their most advanced technology and fully operational 'flying saucers' had escaped from their occupation forces.[142] The US Navy spies discovered this Nazi exodus, according to William Tompkins, and reported on it in their debriefings at Naval Air Station, San Diego up to early 1946.

In a private interview, I asked Tompkins whether the Nazis took everything over to Antarctica at a certain point before the war's end. He replied:

> Yes. And that include everything medical. Okay? Everything in a hospital. Everything in research. Everything in drugs. All of that stuff went down there into Antarctica. And so, when the war

stopped and we got Paperclip, that was only part of what had originally been operating a year before in Germany…. Actually, operatives talked about specific different German people heading up programs. Just they've flew these names off this ledger [sic]. You couldn't believe it. On different programs, even on the medical research side. It went to Russia, and so like [on] a 20:1 or something ratio. We got good ones, but the best ones went to Antarctica and to support programs.[143]

These events revealed by Tompkins are confirmed by Corey Goode based on his own reading of documents in an SSP digital archive describing the German exodus to Antarctica. Goode now concludes that the documents he read during his "20 and back" service were based on Tompkins' earlier briefing packets:

I'm really beginning to believe that a lot of his briefings during the time period – 1942 – were what made it into the database that I was reading on the smart glass pad, you know, like 30 [40] years ago … we were viewing old typeset documents.[144]

Support for Tompkins and Goode's claims of Navy intelligence documents detailing what really occurred at the end of World War II come from a number of sources. Richard Wilson and Sylvia Burns state they saw classified documents while researching their book, "*Secret Treaty: The United States Government and Extra-terrestrial Entities.* This is what they say they discovered about Nazi Germany:

The Germans in the scientific community knew the war was lost as early as 1942. They decided to establish a plan for continuing the dream of the Third Reich despite the war. They decided that the establishment of a separate society founded on Nazi principles of genetic purity was the answer. The development of gravitational technology aided that plan. On February 23, 1945, the newest engines of the Kugelbitz were tested and then extracted from the craft. The Kugelbitz was blown up by SS personnel and the scientists, plans and engines were shipped out of Germany to the South Polar regions, where the Germans had maintained underground construction activity since 1941. Two days later, on February 25, 1943 the underground plant at Khala was closed and all the workers sent to Buchenwald and gassed. The Germans also sent their "aryan elite" children and other elements of their society to the underground base. General Hans Kammler, who disappeared in April 1945, was instrumental in the evacuation operation, as was General Nebe. There, the Germans developed a eugenic society that apparently is limited to a specific number of people. They're still there. Apparently they also maintain technical colonies in South America.[145]

Another compelling source reinforcing Tompkins and Goode's claims is the anonymous CIA agent known as Kewper, who revealed that Nazi Germany's most advanced flying saucer programs had been moved to South America and Antarctica prior to the outbreak of the Second World War.

The British had come up with photographs of the saucer craft in the 1930's, and so we knew Germany had the saucer craft with 'laser guns' on them. Hitler actually sent out all the craft that they had to Argentina and Antarctica apparently to make sure when he started WWII none of them would be captured.[146]

In an interview with Linda Moulton Howe, "Kewper" described some of the Nazi craft that had been relocated from Peenemunde to South America:

> [Howe] You did have a photograph that confirmed the Germans were still flying some of their Peenemunde craft in South America?
> [Kewper] Oh, yeah! The craft with a high center about 12-feet high – they all look like Haunebu II's. Although they could be alien craft as well. But we labeled those photos as being German craft from Argentina. However, on radar, we used to see some of the *real* alien craft come from outer space right down into the Argentina region. We also saw craft come into the Antarctica region from outer space via radar we shared with the British down in the Falkland Islands in the South Atlantic Ocean east of Argentina…. In 1959 to 1960, our unit was separating alien craft from the known German craft by the appearance of the craft. We always found the German craft to be much slower in speed than the alien craft. Some alien craft were tracked from outer space doing something like 30,000 mph![147]

What remained of Nazi Germany's advanced weapons programs was disturbing enough in terms of the overall technological advances achieved by the Nazis in many fields of weapons production.[148] However, discovering that the Germans had removed their most advanced secrets, technology and personnel during the run up to the final defeat of Nazi Germany would have been an outright shock to Allied leaders once they realized what had occurred.[149] Rather than the final months of World War II being the last desperate gamble by a megalomaniac

Nazi leadership that could not accept inevitable defeat, it was in fact a holding action for a methodically well planned extraction of the Nazi's most valuable resources and personnel to well-prepared remote locations in the Antarctic and South America. Kammler's successful negotiations with the Allied forces only involved the *second tier* of advanced technologies developed by the Nazi regime.

The Nazis had sufficient time and resources to prepare for such an exodus given their extensive business links, front companies, and connections with South American governments and companies. The Fraternity, as stated earlier, was a transnational group of industrialists and national elites who had extensive business ties, for whom World War II was little more than a hiccup in their lucrative business relationships. The well-equipped Nazi expeditions to the Antarctic in the pre-war period allowed the Germans to familiarize themselves with the terrain, and lay the foundations for any post-war role to be played by this territory.

After Germany's unconditional surrender on May 8, 1945, Nazi submarine activity in the Antarctic region continued as evidenced by the following report presented by the *Agence France Press* on September 25, 1946:

> [T]he continuous rumors about German U-boat activity in the region of Tierra del Fuego ['Feuerland' in German] between the southernmost tip of Latin America and the continent of Antarctica are based on true happenings. [150]

What compounded this realization by the Allies over the Nazi elite exodus to Antarctica was evidence that the Vril Society had been successful in contacting extraterrestrial races, and an advanced civilization based in the Earth's interior. Nazi cooperation with extraterrestrial life and/or ancient interior Earth civilizations would now be a surreal factor in the Allies pursuing and eradicating the remnants of Nazi Germany which had relocated to distant lands.

The Vril Society's saucer program had achieved operational success well before the Nazi defeat. Orsic had ensured that the Vril Society program was not directly associated with the war effort, so this made it possible for them to move most of their technologies, infrastructure and personnel to safe locations in Antarctica and South America. In a private interview, Tompkins described how Orsic's flying saucer program was allowed to move to Antarctica, and why it was kept separate from the Nazi SS program:

> She [Orsic] had two other girls who she actually knew from when they were just kids. So, the three of them were sort of the principals of initiating all this effort in the 1,442, whatever, people involved in that first build [of the civilian space program]. Those were like hundreds and hundreds of people that were brought into this thing later on and developing all this, but none of them, to my knowledge, went to Antarctica like the pretty girls did. So, it's a massive amount of information. Space transports, different programs. Her involvement with the SS, Michael [the author], she started it, okay, because the Nordics gave her the information originally and this was like a year before anything was picked up with the SS. There were two separate programs going, hers and the SS. And then when the SS finds out about her and everything that she's getting and doing, they confiscated everything that she had, and all the people that were working for her, with her.
>
> And then somebody got smart over there on the SS

side and said, "Hey, this is the second group of extraterrestrials, not just the Nordics, the Reptilians that are helping us, but we got Nordics too." So, they gave everything back to her. Okay? She was allowed, Michael, to — She didn't want anything to be military. And of course, the only thing the SS wanted, The Third Reich, well, was all military, but here were two separate areas found out about.[151]

It's important to keep in mind Orsic's disdain for the Nazi SS and its efforts to militarize space technologies that she had played a key role in initially developing. She was part of a far more pacific German mindset that emphasized cosmic philosophy and peace, based on *correct* use of the Vril Force. Orsic's skillful ability to maintain her own Vril space program in Antarctica would later prove to be pivotal in the development of the Space Brothers social/philosophical movement that erupted in the 1950's.

The large scale movement of Himmler's SS personnel and resources to Antarctic did not happen until very late in the war. The SS advanced weapons projects under Kammler's authority had failed in its last ditch effort to weaponize flying saucers in time to turn the tide of the war. Kammler moved whatever infrastructure and operational flying saucer craft he could before the advancing allied armies reached the top secret SS production centers in Pilsen, Czechoslovakia and elsewhere. This is how author and researcher W.A. Harbinson describes the Nazi SS retreat to Antarctica according to documents he witnessed:

[I]n March 1945, just before the end of the war, two German provision U-boats, U-530 and U-977 were launched from a port on the Baltic Sea. Reportedly they took with them members of the flying saucer research teams, the last of the most vital flying saucer components, the notes and drawings for the saucer, and the designs for gigantic underground complexes and living accommodations based on the remarkable underground factories of Nordhausen in the Harz Mountains. The two U-boats duly reached Neuschwabenland, more correctly known as Queen Maud Land, where they unloaded.[152]

In response to a question about the role of Orsic and the Vril Society in post-war Nazi bases, Goode explains:

> She has obviously made it to the Antarctic Bases/Cities.... I do know that these "Societies" were very much the backbone of what survived the war and who were in control of the facilities along with the group they believed was ET, as well as the [Reptilian] Draco Federation that they allied themselves with.[153]

This is a very revealing statement. The Vril Society and its allies in the Thule and Black Sun societies, had not only successfully maintained their independence from Himmler's SS, but now they were in a leadership position in the Antarctic facilities. The Nazi defeat led to incorporating the remaining portion of Kammler's weaponized flying saucer program into the operational secret space program developed by the Vril, Thule and Black Sun societies, which had gone undetected and undamaged through World War II.

The German alliance's involvement with Draconian extraterrestrials led to the Reptilians helping the Nazis establish their Antarctica bases, according to what Tompkins says he heard during the Navy spies' debriefings:

> The fact that the Germans were given this information from the Reptilians. They set up the

program, they designed the program to support Germany, and they are giving Germany the UFOs.... Now in addition to that, they [Nazis] had, if you want to call them, "consultants", who are Reptilian consultants assisting on all of these different things that it takes to design and build these spacecraft carriers, and propulsion systems. So this is an extremely well developed program and documented like crazy. Getting copies of the documents was hard for them, hard for our spies. This was an open program in the upper level of the SS.[154]

The success of the Vril Society in maintaining its independence from the Nazi SS, and even assuming a leadership position in the breakaway society, shows how powerful Vril, Thule and Black Sun leaders had become in Hitler's totalitarian society. Within their Antarctic bases, along with those in Argentina, the now merged Nazi SS and German Secret Society space programs had fully operational 'saucer ships' that could move around the planet and even fly to a secret moon base. According to reports from a number of military officials aware of the advanced nature of the Nazi's technology, many of the UFOs witnessed in the immediate post war era were fully operational German spacecraft.[155]

Operation Highjump Encounters Nazis & ET Allies

A U.S. effort to locate, take-over and/or destroy the Nazi Antarctica bases occurred with a Naval military expedition led by Rear Admiral Richard Byrd in 1946/47. Byrd's military expedition to Antarctica was named "Operation Highjump", and comprised "4700 military personnel, six helicopters, six Martin PBM flying boats, two seaplane tenders, fifteen other aircraft, thirteen US

Navy support ships and one aircraft carrier; the USS Philippine Sea."[156] Byrd received both classified and unclassified orders, an occurrence which is not unusual for Navy commanders. Yet for the general public, only the unclassified reason for this expedition was released, explaining its purpose as largely scientific in terms of exploring, mapping, and finding locations for future U.S. bases.

Among the unclassified commands issued to Admiral Byrd from the Commander of the US Navy, Admiral Chester Nimitz, were:

a) train personnel and test material in the frigid zones

b) consolidate and extend American sovereignty over the largest practical area of the Antarctic continent

c) to determine the feasibility of establishing and maintaining bases in the Antarctic and to investigate possible base sites

d) to develop techniques for establishing and maintaining air bases on the ice, (with particular attention to the later applicability of such techniques to Greenland)

e) amplify existing knowledge of hydrographic, geographic, geological, meteorological and electromagnetic conditions in the area.[157]

These unclassified scientific reasons for the expedition were a strategic cover for its critical classified purpose. As its size clearly suggested, Byrd's Naval armada was not conducting a scientific mission, but instead a military expedition against a publicly unrevealed secret enemy.

A year earlier during the South Pole summer of 1945/46, according to Tompkins, Admiral Byrd led a secret mission to

Antarctica to negotiate with the leaders of the German colony established there. Further, the mission was unsuccessful, and Byrd returned empty handed. In a private interview, Tompkins elaborated about this event:

> [Q]: You have previously spoken about Admiral Byrd being privately taken to Antarctica in 1945/1946 (the year before Operation Highjump) where he attempted to negotiate with Nazis there. Can you elaborate on what happened and why the negotiations failed?

> [Tompkins]: Okay. Basically, it's kind of simple. He didn't have anything to bargain with. Okay? When they went down there, when he attempted to negotiate, if you want to say that, you know, he's got [Inaudible] with nothing in it. He didn't have any power at all. And so, it's nothing more than an attempt at something that you know was not going to work. He had no bargaining power.[158]

The testimony of Navy Commander Graham Bethune, who says he was the actual pilot of the plane that took Byrd to Antarctica, supports Tompkins' claims. Bethune told Lieutenant Colonel Donald Ware (USAF, ret.) about this secret mission, which was one among the many he flew involving classified UFO/antigravity craft. Ware reported details of their personal conversation as follows:

> Graham Bethune was a friend of mine … He was the pilot who flew Admiral Byrd to Antarctica in 1945. Byrd was sent down there to find out what was going on. He was sent home early. But I think Byrd was able to negotiate U.S. scientists joining the German scientists in their underground base

working on advanced technology.[159]

The information Bethune shared serves as a harbinger of the close cooperation soon to be developed between U.S. and German scientists, and the U.S. aerospace industry with their German partners; all beginning after a fateful agreement was reached by political leaders.

Due to the harsh seasonal conditions after the failed 1945/1946 mission, the next Antarctic summer presented the first feasible opportunity to mount such a large military deployment to the frigid regions of Antarctica. Yet, coming so soon after the end of the Second World War, it is puzzling that such a sizable armada would travel to Antarctica at a time of increasing Cold War tensions, in tandem with the decommissioning of Naval personnel and ships, unless the expedition was sent to militarily deal with some unresolved issues from the war itself. Specifically, to deal with the remnants of the Nazi elite hidden in an underground base or bases established in the pre-war era. Byrd's classified mission was therefore critical in its goal to locate, take-over and/or destroy the Nazi Antarctica bases.

Ironically, before Captain Ritscher's Schwabenland expedition departed for Antarctica on December 17, 1938, Admiral Byrd was invited to Nazi Germany as a guest of honor for its opening ceremony.

> In 1939, before the German Expedition, the only aerial photography undertaken in the Antarctic had been that by the most famous U.S. expeditioner, Richard E Byrd in 1933. Byrd had visited the German Expedition in Hamburg at the invitation of the German Society of Polar Research which had the task of assembling and training the crew of the Expedition. The Society invited Byrd to join the Expedition and he met the crew and was given a tour of the preparations. Byrd, however, declined

the offer, returning to the U S and taking command of the U S Antarctic Service at the request of President Roosevelt until that position was ended in its first year due the approach of World War II.[160]

As mentioned earlier, on July 8, 1939, Admiral Byrd first received orders from President Roosevelt to locate and challenge any Nazi Antarctic bases that fell under the U.S. sphere of influence.[161] The orders were not carried out then, or during the U.S. entry into the war. Finally in 1947, Admiral Byrd led a U.S. Naval Expedition to effectively deal with the bases it found that had been created or 'discovered' during and/or after the 1938 Schwabenland expedition. Arguably, the Nazi Antarctic bases had nine uninterrupted years to prepare for this looming confrontation with the US Navy.

The Byrd mission was scheduled to last up to six months, but ended in a mere eight weeks because it had, according to initial Chilean Press reports; "run into trouble" and there had been "many fatalities[162]". If the true goal of the mission was to locate and eradicate Nazi bases, the press reports and early end to the mission indicate a dismal failure and a rude awakening for the US Navy.

The most commonly cited source for learning the truth behind the early end of Operation Highjump is an interview Admiral Byrd gave to journalist, Lee Van Atta, in Santiago, Chile while traveling back to the U.S. in early March 1947. Van Atta's article has many significant direct quotes from Byrd suggesting he was eager to reveal the truth, so it appears he was later muzzled when he returned to Washington, DC.[163] Van Atta's article deserves close scrutiny of its key passages citing Byrd because these give the most accurate and honest assessment of what really took place in Antarctica.

Van Atta's article first appeared in the March 5, 1947 edition of *El Mercurio* and was titled, "Admiral Richard E. Byrd refers to the strategic importance of the poles" (see Figure 20).

What follows are translated passages from the original Spanish release:

> Admiral Richard E Byrd warned today of the necessity for the United States to adopt protective measures against the possibility of an invasion of the country by hostile aircraft proceeding from the polar regions. The admiral said: 'I do not want to scare anybody but the bitter reality is that in the event of a new war, the United States will be attacked by aircraft flying in from over one or both poles.'"[164]

This specific passage is the source of the original theory proposing that the South and North poles were the locations where a new enemy existed capable of initiating a "new war" against the United States. The reference to a "new war" clearly indicates that the entity involved is both hostile and powerful since it can directly threaten mainland America from the South Pole. Van Atta continues:

> "On the subject of the recently terminated expedition, Byrd said that 'the most important of the observations and discoveries made [were] ... of the present potential situation as it relates to the security of the United States ... I can do no more than warn my countrymen very forcibly that the time has passed when we could take refuge in complete isolation and rest in confidence in the guarantee of security which distance, the oceans and the poles provide.'[165]

Here, Byrd is suggesting that the new enemy possesses advanced aerial craft whose speed and range negated the protection previously enjoyed by the U.S. in terms of vast oceans separating its mainland from Europe and Asia.

After the Soviet collapse in 1991, the KGB released

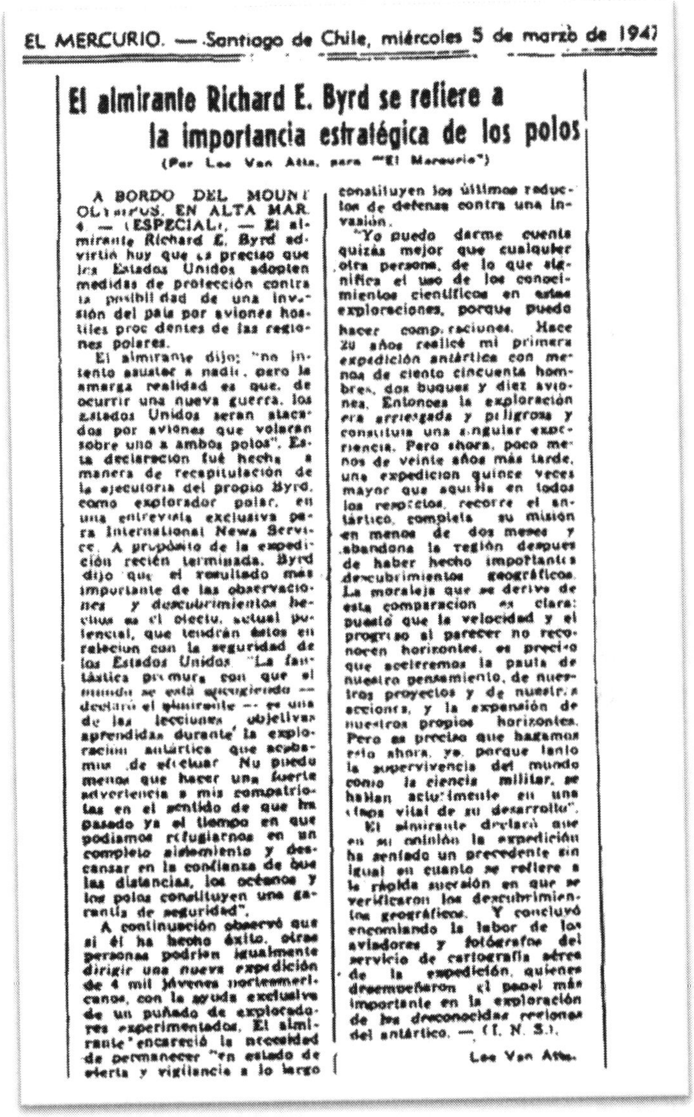

EL MERCURIO. — Santiago de Chile, miércoles 5 de marzo de 1947

El almirante Richard E. Byrd se refiere a la importancia estratégica de los polos

(Por Lee Van Atta, para "El Mercurio")

A BORDO DEL MOUNT OLYMPUS, EN ALTA MAR, 4 — (ESPECIAL). — El almirante Richard E. Byrd advirtió hoy que es preciso que los Estados Unidos adopten medidas de protección contra la posibilidad de una invasión del país por aviones hostiles procedentes de las regiones polares.

El almirante dijo: "no intento asustar a nadie, pero la amarga realidad es que, de ocurrir una nueva guerra, los Estados Unidos serán atacados por aviones que volarán sobre uno o ambos polos". Esta declaración fué hecha a manera de recapitulación de la ejecutoria del propio Byrd, como explorador polar, en una entrevista exclusiva para International News Service. A propósito de la expedición recién terminada, Byrd dijo que el resultado más importante de las observaciones y descubrimientos hechos es el efecto actual potencial, que tendrán éstos en relación con la seguridad de los Estados Unidos. "La fantástica premura con que el mundo se está encogiendo — declaró el almirante — es una de las lecciones objetivas aprendidas durante la exploración antártica que acabamos de efectuar. No puedo menos que hacer una fuerte advertencia a mis compatriotas en el sentido de que ha pasado ya el tiempo en que podíamos refugiarnos en un completo aislamiento y descansar en la confianza de que las distancias, los océanos y los polos constituían una garantía de seguridad".

A continuación observó que si él ha hecho éxito, otras personas podrían igualmente dirigir una nueva expedición de 4 mil jóvenes norteamericanos, con la ayuda exclusiva de un puñado de exploradores experimentados. El almirante encareció la necesidad de permanecer "en estado de alerta y vigilancia a lo largo [...]

[...] constituyen los últimos reductos de defensa contra una invasión.

"Yo puedo darme cuenta quizás mejor que cualquier otra persona, de lo que significa el uso de los conocimientos científicos en estas exploraciones, porque puedo hacer comparaciones. Hace 20 años realicé mi primera expedición antártica con menos de ciento cincuenta hombres, dos buques y diez aviones. Entonces la exploración era arriesgada y peligrosa y constituía una singular experiencia. Pero ahora, poco menos de veinte años más tarde, una expedición quince veces mayor que aquella en todos los respectos, recorre el antártico, completa su misión en menos de dos meses y abandona la región después de haber hecho importantes descubrimientos geográficos. La moraleja que se deriva de esta comparación es clara: puesto que la velocidad y el progreso al parecer no reconocen horizontes, es preciso que aceleremos la pauta de nuestro pensamiento, de nuestros proyectos y de nuestras acciones, y la expansión de nuestros propios horizontes. Pero es preciso que hagamos esto ahora, yo, porque tanto la supervivencia del mundo como la ciencia militar, se hallan actualmente en una etapa vital de su desarrollo".

El almirante declaró que en su opinión la expedición ha sentado un precedente sin igual en cuanto se refiere a la rápida sucesión en que se verificaron los descubrimientos geográficos. Y concluyó encomiando la labor de los aviadores y fotógrafos del servicio de cartografía aérea de la expedición, quienes desempeñaron el papel más importante en la exploración de las desconocidas regiones del antártico. — (I. N. S.).

Lee Van Atta.

Figure 23. Lee Van Atta article on Operation Highjump

previously classified files that cast light on the mysterious Byrd led Naval expedition to Antarctica. This release included a 2006 Russian documentary which made public for the first time a 1947 secret Soviet intelligence report commissioned by Joseph Stalin on *Task Force 68's* [official name for Operation Highjump] mission to Antarctica.[166] The intelligence report, gathered from Soviet spies embedded in the U.S., revealed that the US Navy had sent the military expedition to find and destroy one or more hidden Nazi bases. On the way, they encountered a mysterious UFO force that attacked the military expedition, destroying several ships and a significant number of planes. Indeed, Operation Highjump had suffered "many casualties" as stated in the initial press reports from Chile. While there is a possibility this report resulted from U.S. disinformation fed to a known Soviet mole, the more likely explanation is that the report exposes the first known historical incident involving a battle between U.S. naval forces and an unknown UFO force stationed in or near Antarctica.

In the Soviet intelligence report, never before released testimony by two US Navy servicemen assigned to Operation Highjump was revealed. An article in *New Dawn* by Frank Joseph gives a detailed analysis of the two eyewitness accounts (only the latter of which is mentioned in the 2006 Russian documentary). John P. Szehwach, a radioman stationed on the USS Brownson, gave testimony of how UFOs appeared dramatically out of the ocean depths. On January 17, 1947 at 0700 hours, Szehwach says:

> I and my shipmates in the pilothouse port side observed for several minutes the bright lights that ascended about 45 degrees into the sky very quickly... We couldn't i.d., the lights, because our radar was limited to 250 miles in a straight line.[167]

Over the next several weeks, according to the Soviet report, the UFOs flew close over the U.S. naval flotilla that fired on them, which they, in turn retaliated with deadly effects. According to the other eyewitness, Lieutenant John Sayerson, a flying boat pilot:

> The thing shot vertically out of the water at tremendous velocity, as though pursued by the devil, and flew between the masts [of the ship] at such a high speed that the radio antenna oscillated back and forth in its turbulence. An aircraft [Martin flying-boat] from the Currituck that took off just a few moments later was struck with an unknown type of ray from the object, and almost instantly crashed into the sea near our vessel.... About ten miles away, the torpedo-boat Maddox burst into flames and began to sink... Having personally witnessed this attack by the object that flew out of the sea, all I can say is, it was frightening.[168]

There is a major problem with Sayerson's quote however. There has been no torpedo boat named Maddox in the US Navy.[169] In the Russian documentary, the incident described by Sayerson (misspelt Sireson) refers instead to the destroyer "Murdoch". There was, however, no destroyer named "Murdoch" active in the U.S. Fleet in 1947. Instead there was a destroyer named "Maddox" (DD-731), but it did not serve in Operation Highjump. In fact, the USS Maddox was the destroyer fired upon in the Gulf of Tonkin incident of 1964.[170]

According to Frank Joseph, the USS Maddox was "either a torpedo boat, or torpedo-carrying destroyer". He also explains what may have happened to the Maddox mentioned in the Soviet report:

> A USS Maddox was indeed sunk by enemy action, but five years earlier by a German dive-bomber during the Allied invasion of Sicily. Actually there were at least three American destroyers known by

that name (DD-168, DD-622 and DD-731) all of them contemporaneous. The U.S. Navy has long been notorious for falsifying the identity of its ships and re writing their histories if they embarrass official policy.... So too, the "Maddox" cited by Soviet espionage was similarly consigned to an official memory hole.[171]

If Joseph is correct, then it is very possible that a USS Maddox was destroyed during Operation Highjump, and the US Navy changed official records to hide this fact. An alternative explanation is that the 1947 Soviet report contained U.S. orchestrated disinformation conveyed to Soviet authorities by a Soviet mole known and cleverly used by the U.S. intelligence community. Though plausible, this explanation is highly unlikely, given that the U.S. and USSR were still allies at the time of Operation Highjump, and had a common interest in finding and destroying any hidden Nazi base(s) in the South Atlantic.

It is clear that the best the US Navy could muster was no match for the small, but well-armed German elite that survived the war in their remote Antarctic location. The possibility that the Germans were receiving assistance from a technologically advanced ally in fending off the U.S. attack cannot be discounted. The presence of Orsic and other Vril Society members suggest that the Nazis had an esoteric means of communicating with its allies, who were extraterrestrial and/or an advanced subterranean civilization.[172]

This scenario is supported by the testimony given by Kewper who stated he saw documents and received briefings about the fate of Operation Highjump:

In 1946-1947, the United States had a scientific mission to Antarctica under Admiral Byrd and we had military interaction there with aliens and their saucer craft, like a mini-war. We lost all of our aircraft.[173]

Subsequent reports of extensive UFO activity in the Antarctic region are confirmation that it was being used as a base of operations by a new enemy threat, the one that had unnerved Admiral Byrd and, according to his quote in the Chilean press report; could fly from "pole to pole".[174]

Corey Goode affirms the Nazis had help in establishing and protecting the three Antarctic bases:

> There was help from the Draco Federation as well as a group that the Nazi's were led to believe were ET's (referred to as "Arianni" or "Aryans", sometimes called "Nordics") but were actually an Ancient Earth Human Break Away Civilization that had developed a Space Program (referred to as "The Silver Fleet") and created vast bases below the Himalayan Mountains (largest in Tibet and called the system Agartha) and a few other regions.[175]

Here, Goode is here referring to the same "Arianni" that Admiral Byrd wrote about in his posthumously published diaries, whose authenticity is still disputed. These alleged diaries contain Byrd's account of what he encountered when his plane was lost for several hours during a mapping expedition over the South Pole region (an incident which is well documented). Byrd's description of a meeting with a person he calls a Master is worth quoting in full given Goode's information:

> My thoughts are interrupted in a cordial manner by a warm rich voice of melodious quality,
>
> 'I bid you welcome to our domain, Admiral.'

I see a man with delicate features and with the etching of years upon his face. He is seated at a long table. He motions me to sit down in one of the chairs. After I am seated, he places his fingertips together and smiles. He speaks softly again, and conveys the following:

> 'We have let you enter here because you are of noble character and well-known on the Surface World, Admiral.'

Surface World, I half-gasp under my breath!

> 'Yes,' the Master replies with a smile, 'you are in the domain of the Arianni, the Inner World of the Earth. We shall not long delay your mission, and you will be safely escorted back to the surface and for a distance beyond. But now, Admiral, I shall tell you why you have been summoned here. Our interest rightly begins just after your race exploded the first atomic bombs over Hiroshima and Nagasaki, Japan. It was at that alarming time we sent our flying machines, the "Flugelrads", to your surface world to investigate what your race had done. That is, of course, past history now, my dear Admiral, but I must continue on.
>
> You see, we have never interfered before in your race's wars, and barbarity, but now we must, for you have learned to tamper with a certain power that is not for man, namely, that of atomic energy. Our emissaries have already delivered messages to the powers of your world, and yet they do not heed. Now you have

been chosen to be witness here that our world does exist.

You see, our Culture and Science is many thousands of years beyond your race, Admiral.'

I interrupted, 'But what does this have to do with me, Sir?' The Master's eyes seemed to penetrate deeply into my mind, and after studying me for a few moments he replied,

'Your race has now reached the point of no return, for there are those among you who would destroy your very world rather than relinquish their power as they know it...'

I nodded, and the Master continued,

'In 1945 and afterward, we tried to contact your race, but our efforts were met with hostility, our Flugelrads were fired upon. Yes, even pursued with malice and animosity by your fighter planes. So, now, I say to you, my son, there is a great storm gathering in your world, a black fury that will not spend itself for many years. There will be no answer in your arms, there will be no safety in your science.

It may rage on until every flower of your culture is trampled, and all human things are leveled in vast chaos. Your recent war was only a prelude of what is yet to come for your race. We here see it more clearly with each hour... do you say I am mistaken?'

'No,' I answer, 'it happened once before, the dark ages came and they lasted for more than five hundred years.'

'Yes, my son,' replied the Master, 'the dark ages that will come now for your race will cover the Earth like a pall, but I believe that some of your race will live through the storm, beyond that, I cannot say.

We see at a great distance a new world stirring from the ruins of your race, seeking its lost and legendary treasures, and they will be here, my son, safe in our keeping. When that time arrives, we shall come forward again to help revive your culture and your race.

Perhaps, by then, you will have learned the futility of war and its strife...and after that time, certain of your culture and science will be returned for your race to begin anew. You, my son, are to return to the Surface World with this message....'[176]

Based on the above dialogue allegedly written by Admiral Byrd, it would appear that the German secret societies were allowed to use Antarctica as a refuge, perhaps because they were escaping the war in Europe, and were followers of the Arianni – the remnants of the legendary Hyperborian civilization which represented the core of Thule Society beliefs. Hence, it is plausible that the Arianni helped the German flying saucer fleets defend themselves against the US Navy's Operation Highjump, just as Goode claims.

Operation Paperclip Facilitates Secret Negotiations with Antarctic Colony

Operation Paperclip was a highly classified military

program aimed at locating the most advanced technologies developed by the Nazis, along with the most talented German scientists, and then bring these resources to the U.S. in order to kickstart its own advanced aerospace and rocket industries. Paperclip was overseen by the Joint Intelligence Objectives Agency (JIOA); a multiagency organization led primarily by the US Army and Navy. Earlier, Navy Secretary James Forrestal had requested President Roosevelt create the JIOA due to Naval intelligence about the existence of advanced Nazi technology programs. This information was gained through multiple sources, including Admiral Rico Botta's spies working out of Naval Air Station, San Diego. Tompkins describes Forrestal's and the Navy's goals for Operation Paperclip:

> At the end of World War II, Naval Intelligence operators (spies) penetrated virtually every German secret weapons, advance system, rockets, aircraft, UFOs and heavy water [projects] in the country. They located the individuals in these facilities, and they were tagged. When the hostilities ceased, the Naval Intelligence and additional intelligence officers went straight into these locations and removed not only the research scientists, but their documentation, and as much of the weapons system as they could. They were all brought to the United States in what was called Project Paperclip.[177]

In July/August 1945, Forrestal travelled to occupied Germany to see first-hand how Operation Paperclip was proceeding by visiting Navy and Army facilities to view some of the advanced Nazi technologies captured by the U.S. military. Forrestal would have known about Kammler's successful negotiations with Allen Dulles and the Office of Strategic Services over these Nazi technologies. Thus, Forrestal was likely helping the US Navy decide which of the captured Nazi technologies would be worth exploiting for future use.

The development of advanced Nazi submarines was of particular interest to the US Navy because these vessels shared some of the construction techniques used in building flying saucers and the larger cigar-shaped craft capable of travelling into Earth's orbit and beyond. Years later, according to Tompkins, converted nuclear submarines became the first antigravity spacecraft developed by the US Navy, which were used for deep space missions.[178] During his time at Douglas Aircraft Company's think tank, Advanced Design, Tompkins says he worked on this conversion:

> And at Douglas, in the secret think tank, we were looking at every type of space vehicle we would need to go out into the galaxy, then our US Navy. So submarines came up and we discussed this. We said, "That's the easiest thing, the quickest way, we can get out there. We'll just take a regular Navy submarine, pull out the whole nuclear system, put in the anti-gravitational system, and we're going to use these right away."[179]

After the failure of Operation Highjump, the role of German scientists recruited as a part of Operation Paperclip took on an added dimension. Not only did they assist U.S. scientists in understanding the advanced rocketry used by Nazi Germany, but they were used as intermediaries for secret negotiations with the German Antarctica colony. The U.S. authorities wanted to know how successful the Nazis were in their studies to understand the extraterrestrial technologies, since serious U.S. studies were only just beginning. On this subject, Corey Goode writes:

> [A]fter the failed Operation High Jump Mission the Operation Paperclip Scientists were asked to

negotiate meetings. The Nazi Breakaway group knew that the Americans had recovered crashed craft from several different species of off world visitors that were so far advanced that they were getting nowhere with the reverse engineering of their technology.[180]

The extreme importance of tapping into the knowledge base of the Antarctica German colony meant that Paperclip's German scientists were needed for their skills during secret negotiations that opened the door to these technologies. Therefore, many of them who were previously classified as ardent Nazis were now brazenly reclassified so they could both enter the U.S., and begin work in sensitive U.S. facilities. Jim Marrs writes:

> [O]fficers of the Joint Intelligence Objectives Agency (JIOA) who managed Project Paperclip soon began receiving security reports … regarding the Germans recruited into the program. All reports on these men had been altered from a determination of "ardent Nazi" to read "not an ardent Nazi" … even Wernher von Braun, who in 1947 had been described as "a potential security threat" by the military governor, was reassessed only months later in a report stating, "he may not constitute a security threat to the U.S." Likewise, von Braun's brother, Magnus, who had been declared a "dangerous German Nazi" by counterintelligence officers, was brought to America and his pro-Nazi record expunged. "Serious allegations of crimes not only were expunged from the records, but were never even investigated."[181]

Operation Paperclip scientists were granted security clearances more easily than their U.S. peers, which gave them a distinct advantage when it came to applying for jobs offered by leading aerospace companies, as Marrs explains:

> German scientists could obtain necessary security clearances more easily than could American scientists. Defense contractors looking for new employees to work on classified projects found this aspect of National Interest to be particularly advantageous. By 1957, more than sixty companies were listed on JIOA's rosters, including Lockheed, W. R. Grace and Company, CBS, Laboratories, and Martin Marietta. [182]

William Tompkins witnessed the level of penetration achieved by the Germans entering the U.S. aerospace industry via Operation Paperclip. He says that Douglas Aircraft Company was one of the few not taken over due to it having its own top German scientist, Dr. Wolfgang Klemperer, who came aboard prior to the start of WWII:

> Douglas was not really taken over like the three other aircraft companies were because Douglas had a German guy [Klemperer], a German scientist/engineer, who had been working at Douglas since 1934, I guess.

> And so, he had come over and he was not part of the Nazi party or any of that political stuff at all, but he was a brilliant scientist. And so, he got the package [Tompkins briefing packages from Naval Air Station, San Diego], too, but he wasn't like this under-the-table type of thing that everybody else got. [183]

Notably, Operation Paperclip officially continued up to

1990, at times expanding its recruitment for the U.S. aerospace industry to acquire the most talented German scientists who had ended up working in other countries:

> Another program, code-named simply Project 63, was designed specifically to get German scientists out of Europe and away from the Soviets. "Most went to work for universities or defense contractors, not the U.S. government".... Thus the American taxpayer footed the bill for a project to help former Nazis obtain jobs with Lockheed, Martin Marietta, North American Aviation, and other defense contractors during a time when many American engineers in the aircraft industry were being laid off".... Paperclip again began to grow. Specialists were imported from Germany, Austria, and other countries under Project 63 and National Interest and gained positions at many universities and defense contractors, including Duke University, RCA, Bell Laboratories, Douglas Aircraft, and Martin Marietta.[184]

After Byrd's stunning Antarctic defeat, the Germans used their hidden bases and advanced flying saucer technologies to pressure both the Truman and Eisenhower administrations into accepting secret deals. This culminated in the 1952 Washington Flyover, which had the effect of accelerating the negotiations between Washington, DC and the Antarctica German colony.

1952 Washington, DC Flyover

Tompkins revealed in various interviews that the flyover was done by the Antarctica based Germans. In a private conversation, I asked Tompkins if Antarctica based Nazi spacecraft flew over the U.S. in the summer of 1952. He replied:

> It's of course a Yes. Some had the swastika on them and some had the German cross, but they were all extraterrestrial type vehicles, okay, which the Germans had re-engineered, reversed, whatever, and were putting it in production. So, those vehicles were not powered by extraterrestrials or extraterrestrial vehicles. These were German vehicles that had been given to Germany by the Reptilians, but these were production vehicles out of the production facilities in Antarctica.[185]

Similarly, Goode says in an interview:

> They had also received intelligence from their Paperclip spies that the Americans had implemented an Executive Order making the existence of alien life the most classified subject on the planet. The reason being that the development and release of free energy would quickly destroy the oil trade, and soon thereafter the entire Babylonian Money Magic Slave System that all Elites use to control the masses. The NAZI's used this to their advantage in some very public sorties over Washington, DC and highly Secret Atomic Warfare Bases to mention a few. Eisenhower finally relented and signed a treaty with them (and a few other groups, both ET and Ancient Civilizations pretending to be ET).[186]

Figure 24. Photo of 1952 Washington, DC Flyover

Tompkins and Goode's controversial claims are corroborated by Clark McCelland, who worked for 34 years with NASA and finished his career as a Spacecraft Operator. In the August 3, 2015 installment of his book, *The Stargate Chronicles,* McClelland writes:

> The over flights of advance very swift crafts over Washington, DC were these German advanced aircraft that totally out flew American advanced crafts. On July 12, 1952, President Truman observed several of the UFOs and was completely amazed by their capabilities of outmaneuvering the USAF ... advanced Jet fighter ... [Lockheed F-94 Starfire]. USA jets sent up to bring one down. None could fly the speed of the German Saucers.[187]

McClelland, also describes the role of Nazi scientists who had fled to Antarctica in relation to the 1952 Washington Flyover:

Because I worked with the German Scientists that were brought to the USA by Dr. Werner von Braun in 1946/7. Several told me that WWII German Scientists by the many thousands escaped from Germany near the fall of Germany in WW II. They boarded advanced submarines in the Baltic Sea. They were all taken to the South Pole base located underground, in Antarctica. Some called it Hitler's Shangri La. Those scientists created advanced anti-gravity craft that were flying in our air space for many years. And still are. They were observed over Washington, DC in 1952 by President Harry S. Truman. Yes, we did not have any aircraft that could stop these German planes from flying over our national capital in 1952. So German scientific expertise was again showing the USA who was boss.[188]

In addition to brute displays of technological superiority by the surviving Nazi regime, Goode, Tompkins and McClelland claim that there was an extensive infiltration of the military industrial complex by Nazi sympathizers. Among the thousands of former Nazi scientists and technicians who were part of Operation Paperclip, there were assets from the Antarctic based German breakaway group whose job was to infiltrate the U.S. Space program and military industrial complex. The latter was well on track to establishing its own "breakaway civilization", which according to Goode was infiltrated and coopted by Nazi assets who were promoted through the ranks:

When both Truman and Eisenhower signed treaties with the NAZI Break Away Civilization/Societies, it was then that the already well placed Operation Paperclip Operatives (in Military, Corporate Industry, Intelligence and established Secret and

Public Space Programs) easily slid into more powerful and influential positions over the massive industrial complex of the USA that they coveted to expand their operations in space....[189]

After the 1952 Washington Flyover, negotiations picked up and several grueling years later a deal was finally reached with the Antarctica German colony, which had expanded with the help of "The Fraternity", with its industrialists and national elites from around the world who assisted German companies operating in Antarctica. The pivotal benchmark deal was agreed upon at Holloman AFB in February 1955.

CHAPTER 5

The Secret Agreement:
U.S. Military-Industrial Complex Collaboration in Antarctica

President Eisenhower's Secret Meeting at Holloman AFB

On February 10, 1955, President Dwight D. Eisenhower flew on Air Force One from Washington, DC to Thomasville, Georgia for a "hunting vacation". He was accompanied by a chartered plane filled with the Press. Later that afternoon after landing, Eisenhower disappeared from Press view for the next 36 hours. James Hagerty, his press secretary, told the press that Ike and his valet were "treating a case of the sniffles..."[190] In reality, circumstantial and testimonial evidence reveal that he secretly traveled to Holloman Air Force Base, New Mexico, on February 11 to meet representatives of the German-Reptilian alliance, who were behind the 1952 Washington UFO flyover.

The Holloman meeting occurred almost a year to the day after a February 20, 1954 meeting at Edwards Base where human-looking "Nordic" extraterrestrials attempted to dissuade the Eisenhower administration from developing thermonuclear weapons, and warned about the Reptilians and their German allies.[191] The Nordic outreach was rejected, thereby opening the door for agreements to be reached with the Reptilians/Germans.

Once again, there is both circumstantial and testimonial evidence that Eisenhower did secretly travel to Holloman AFB to finalize an agreement with the Antarctic based Germans who were working with a group of Reptilian/Gray extraterrestrials. The first testimony comes from UFO researcher Art Campbell from his 2007 interview with a security guard for Air Force One (aka Columbine III) who confirmed that it secretly left Spence Air Force Base, Georgia, on February 11 at 4 am with Eisenhower on board. Campbell described his conversation with the security guard:

> Then he said... "I do recall one trip down to south Georgia (he wasn't on this one) where there were a dozen or so going to this tiny little town." He went on to say that plane crew did not ask any questions, but they learned why the following day. About 3:00 a.m. they had gotten word that the president would be leaving in an hour. "We were always ready for this kind of thing, and sure enough, the plane left one hour later." He said about a half hour before the plane left, two Air Force cars pulled up and six agents came on board. They had apparently been booked into a nearby motel somewhere for a day or so. The other agents in the little town bustled around in their darkened vehicles, indicating that the president was there. No one noticed when the president returned late at night a day or so later, and no one ever knew he had left.[192]

More testimonial evidence comes from former USAF medic, Bill Kirklin, who from March 1, 1954 until August 5, 1955 was stationed at the Holloman Air Force base hospital. He claims he received prior notification of an impending visit by Eisenhower in February 1955. Kirklin wrote:

… we heard that the president was coming to Holloman. I knew there was going to be an honor parade for him. Captain Reiner asked me if I wanted to participate in the parade. I said, "No." He said, "Fine. You will be on duty." The Parade was scheduled for early in the morning. The day before it was to take place it was called off.[193]

At the end of the day of Eisenhower's visit, Kirklin reports that he saw Air Force One leaving the base and fly over a restricted area:

"After work I was in my barracks room when I was called out to see Air Force One fly overhead. It flew over the residential area of the base. This is a NO FLYING zone for all military aircraft. Only the President could get away with it."[194]

The above is solid circumstantial evidence that Eisenhower was not recuperating in Georgia as his press secretary claimed. Instead, Eisenhower was secretly over 2000 miles away at Holloman Air Force Base. Various aspects of Eisenhower's actual encounter with UFOs and their occupants are also revealed by several first hand witnesses.

Kirklin claims that he heard a number of people commenting on the flying saucers that had arrived at Holloman AFB during Eisenhower's visit. He says one was his colleague, Dorsey, who told him:

"Kirklin, did you see the disc hovering over the flight line?"

"No." I am thinking something small you hold in your hand like a discus as the only craft I knew capable of hovering were the choppers and the Navy's hovercraft. There weren't that many helicopters around Holloman. "What's it made of?" I am thinking of a wooden disc with a steel edge. "Looks like polished stainless steel or aluminum. You know just bright metallic and shiny."

I asked, "How big is it?"

"Twenty to Thirty feet in diameter. Do you want to see it?"

"Sure. But with my luck it wouldn't be there."

Dorsey replied. "It was there when I took my wife to the Commissary and it was there when we got out thirty minutes later. Go out to the front of the hospital and take a look."[195]

If Kirklin's account of what his colleague saw is correct, then at least one flying saucer was hovering over the flight line of the base for at least 30 minutes during Eisenhower's visit.

Later Kirklin went to the mess hall, and says he overheard the following conversation:

On the way back I followed two pilots. The one on the left was in Khakis, the one on the right in winter Blues. I followed them and listened to their conversation.

Left: " Why the Blues?"

Right: "I'm the Officer of the Day, I was at Base Ops when Air Force 1 came in. Did you see it?"

L. "Yes. It's a big bird isn't it?"

R "Yes. They landed and turned around and stayed on the active runway. We turned off the RADAR and waited."

L. "Why did you turn off the RADAR?"

R. "Because we were told to. I think the one at Roswell that came down was hit by Doppler Radar. It was one of the first installations to have it in the U.S. Anyway, they came in low over the mountains, across the Proving Grounds.

Interrupted by L. " I heard there were three and one landed at the Monument."

R "One might have stayed at the Monument. I didn't see it. I only saw two. One hovered overhead like it was protecting the other one. The other one landed on the active [runway] in front of his plane. He got out of his plane and went towards it. A door opened and he went inside for forty or forty-five minutes."

L. "Could you see? Were they Grays?"

R. "1 don't know. They might have been. I couldn't see them. I didn't have binoculars." ...

L. "Do you think these were the same ones that were in Palmdale last year?"

R. "They might have been." ...

R. "It might have been. I just don't know."

L. "Did you see them when he came out?"

R. "No. They stayed inside. He shook hands with them and went back to his plane."

Importantly, these pilots reveal that Eisenhower disembarked from Air Force One and met with the occupants of the flying saucer that landed at the end of the flight line for at least 45 minutes. It's also significant that the pilots refer to the Palmdale (Edwards AFB) meeting in February 1954. Perhaps most noteworthy is the handshakes at the end of the meeting. This is evidence that an understanding or an agreement had been reached. As I will shortly show, indeed this is what happened at Holloman on February 11, 1955.

The family of a base electrician, who worked at Holloman and witnessed a flying saucer approach the area where Air Force One was positioned, later contacted Art Campbell. He was given a letter from the electrician explaining what happened:

> So the day the President came we went out in the truck to a job where we were replacing some wire down the flight line.... So we heard the President's plane in the morning lining up for an approach and watched it land on the far runway. So we waited for it to taxi over to the flight line so we could see him, but we didn't hear it anymore. It had shut down somewhere out there ... one of the men ... said he can see out there from that pole over there, so why don't one of us go up the pole and see where the plane is? Well I had one of my climbers on and ... started up with my back to the sun, a safety measure, which also put my back to the runway where we thought his Connie was. Connie was a nickname for the big Constellation the President flew.... A few minutes later ... I could not believe what I saw. There was this pie tin like thing coming at me about 150 feet away. I thought it was remote controlled or something. 25 to 30 feet across and I started down the pole as fast as I could go.... While I was running towards the big hangar I looked back and it had stopped and it was just sitting there.[196]

The electrician's story is very revealing since it is rare first hand testimony that Eisenhower's Air Force One had landed at the end of the flight line, and was waiting to meet up with a flying saucer.

Another direct witness is an airman whose plane was delayed at Holloman AFB on the morning of Eisenhower's arrival at Holloman. The airman, Staff Sergeant Wykoff, reveals what happened in an interview:

> We had to haul a load of stuff down there. Parts that they needed, and the runway is like this. I [had] never seen anything like that before. And anyway as we were there we saw Air Force One come in, and we didn't know who it was. And then an officer comes around and said you can't leave. The pilot said we have to leave. And he said well President Eisenhower is here and you can't leave the field until he's gone. And I said, you can hear it over loud speakers, but it didn't do any good. I would have liked to have seen him... We didn't have the clearance to go into the mess hall and one of the other officers, a higher ranking officer came [over] to us and said would you like to go in and eat, and listen to his speech. And most of us said yes, because I'd like to see him. I didn't get to shake his hand or anything, because I didn't have the right badge, clearance to go in, but they did let us go in at the very end and we ate and listened to his speech.[197]

Once again, Wykoff is a rare first hand witness who actually observed that Eisenhower had secretly arrived at Holloman AFB base to perform some classified activities. The classified nature of Eisenhower's activities is revealed in the following recollections by Bill Kirklin of a conversation he took part in at the base hospital with a doctor displaying the rank of Captain and Lt. Colonel:

After supper I saw the lights that were still on in the Flight Surgeon's Office and went over to turn them off. I saw Dr Reiner talking to a Lt. Col... The Colonel was talking. "He was at the supply hanger. I was there in the front with him and some others. I was on the stage. There was standing room only with 225 men in the hanger."

Dr. R. "I heard that he was at the base theater."

Lt. Col. "He might have been. He only spoke for a few minutes. Then the base Commander spoke for about twenty minutes. He had plenty of time to go to the base theater and get back."

Dr. R. "How many did he talk to?"

Lt. C. "I was there for two sessions standing room only. 225 each time. There might have been another session but I wasn't there if he spoke then."

I asked, "Who spoke?"

Lt. Col. "The Commander in Chief"

I said, "The President ... "What did he talk about?"

Lt. Col. "It's classified."

"Confidential?"

"Higher."

"Secret?"

"Higher."

I said, "Oh."

Lt. Col. "What do you mean by 'Oh?'"

"It is none of my business. I am only cleared to secret."

Lt. Col. "I would not say that if I were you."

If Kirklin's recollection is correct, then the activities that Eisenhower performed at Holloman AFB on February 11, 1955 were classified above Top Secret. Base personnel were taken into a large hangar and debriefed in groups of 225.

Further proof comes from Clark McClelland who extensively interacted with German scientists such as Werner Von Braun, Kurt Debus and many others while he worked at NASA facilities in Florida. McClelland describes what Dr. Ernst Steinhoff, another German Paperclip scientist, told him about the Holloman incident. At the time of the secret meeting, Steinhoff was visiting Holloman because of his pending transfer there.

> Dr. Steinhoff ... did say, [he] was there during what was called a surprise visit by USA President Eisenhower who flew in with no early notice of those I spoke with that worked there.... The base United States Air Force Officer that managed Holloman, Colonel Sharp, did have prior knowledge of his arrival but not all who worked there.... It was a big surprise to all others who saw his large plane land... [198]

This is consistent with what Kirklin and others have said about the Holloman AFB landing involving Eisenhower.

McClelland went on to say that Steinhoff told him that the meeting involved Antarctic based Germans:

> It was a German Flying Saucer that he [Steinhoff] and others saw at this base. President Eisenhower, being from German heritage realized that when he

was met by a German officer as he boarded that Saucer. The President then realized why none came forth to greet him as he entered that German advanced flying machine.[199]

McClelland also points out the connection between the craft that landed at Holloman, and those that overflew Washington, DC in July 1952:

> I recall something I heard Dr. Kurt Debus say to Dr. Knoth, the Senior Scientist at KSC [Kennedy Space Center], as I entered his office one day. He was speaking of a V-7 craft and my entrance startled both of them. I apologized for walking in on them. Dr. Debus said it was "OK, Clark" to me. Later, I discovered through another German Scientist that the V-7 was the code name for a German Saucer shaped craft that was developed below the South Pole Ice Cap. The same type that overflew Washington, DC and startled President Truman and the Pentagon Chiefs in 1952.[200]

Finally, William Tompkins confirmed the Holloman AFB meeting involving President Eisenhower in his response to questions I asked in a private interview:

> Salla: [#26] Did Antarctica based Nazi spacecraft land at Holloman Air Force base in February 1955 to meet with President Eisenhower?

> Tompkins: And I have to say on #26 that's a Yes also. Essentially, Eisenhower accepted like a defeat in the war without it actually being that. There was really nothing he could do, like it's really one-sided situation.

Salla: So, basically, it was a negotiated surrender.

Tompkins: Yeah. Like he lost that war. He surrendered. [201]

Secret Agreement allows Antarctic based Germans to infiltrate U.S. Military-Industrial Complex

After agreements were reached in 1955, both the U.S. and Germans began a race to see who could infiltrate the other more quickly, but the odds were stacked against the U.S. intelligence community. Corey Goode explains what happened:

> After the treaty was signed and the joint Secret Space Programs began in earnest, things quickly got out of hand and the Nazi Break Away group won the race to infiltrate and take over the other side. They soon controlled every aspect of the U.S. from the Financial System, The Military Industrial Complex, and soon after, all three branches of the government itself. [202]

Goode describes the takeover as a silent coup by the breakaway German regime which had succeeded in establishing its Fourth Reich:

> During the 1950's and thereafter, they had successfully infiltrated and subverted the Military Industrial Complex and major Corporate heads, they had effectively won control of the direction of not only the Break Away Civilization Programs but also the mainstream government and financial system. It was a very effective and silent coup that gutted what was once the American Republic and

turned it too into a Corporate Entity with each of us being "Assets" with our very own serial numbers. This plan was in action far before World War One by various secret societies who controlled the financial system and as many know financed both sides of the wars.[203]

Goode's claims are supported by Tompkins' recollections of how thousands of Nazi scientists were secretly brought in under Paperclip to quietly subvert the U.S. aerospace industry:

Some of the companies got one or two top German people into the company. Others got 20 into the company and virtually every organization in the company.... Not just scientists. Weird! And there wasn't like a 160 of these fellows that came over. These are like thousands and thousands and thousands of them. Okay? And not necessarily the best, but rather were brilliant. So, virtually, every company that handles any kind of army, navy, aircraft kind of group and medical got loaded with Germans. And some of them could hardly speak English, but they came in anyway and they physically turned over everything in the advance design areas. And so, yeah, it was takeover of your country's aerospace.[204]

Further first hand evidence of a silent coup or take over having taken place by the Fourth Reich is found in the recollections of Clark McClelland, who reported seeing Hans Kammler at NASA in the mid-1960's. McClelland says that he met with Kammler at the office of the Director of the Kennedy Space Center, while Kurt Debus was the Director [1962-1974]:

I opened his office door and saw two people I had

not seen at KSC. He introduced me to both men. He only gave me their first names during the introduction.... One was introduced to me as Sigfried and the other was introduced as Hans.... Both had the look of Nazi High Command Officers.... Today I am certain of who these two men were. I eventually learned from other German scientists that one of them me was Siegfried Knemeyer. He was a very high ranking Nazi Oberst Officer in the Luftwaffe... The other man was difficult to recognize until I saw an older photo of him after he had later entered the USA. He was in my opinion Heinz (Hans) Kammler.... There were rumors after WWII that Kammler had made a deal with General George Patton to turn over German Top Secret technology for his support in getting Kammler into the USA. That may have actually happened. I personally believe it did happen.[205]

Due to the secret deal reached between the Eisenhower administration and the Fourth Reich, German scientists brought in under Operation Paperclip were not allowed to be under Army surveillance, thus making them ideal assets for the Fourth Reich, as Jim Marrs explains:

Imported Nazis had every opportunity to pass national security information out of the country.... [T]here was no further army surveillance over the Nazi Paperclip specialists after just four months of their signing a contract with the U.S. government.[206]

There is abundant evidence that Paperclip scientists were spying for the Antarctic based German colony, but could not be stopped due to a secret agreement with the Fourth Reich according to

Marrs:

> Incidents of information being passed out of Paperclip were presented to authorities, yet nothing was done. A Fort Bliss businessman reported Paperclip engineer Hans Lindenmayr to the FBI, claiming the German had been using his business address as an illegal letter drop. According to Hunt, at least three other Nazis maintained illegal mail drops in El Paso, "where they received money from foreign or unknown sources and coded messages from South America." It was also learned that many Paperclip Nazis received cash from foreign sources. "Neither Army CIC nor FBI agents knew where that money came from, and by all appearances, no one cared to know how more than a third of the Paperclip group suddenly were able to buy expensive cars," noted Hunt. [207]

Many Nazi Paperclip scientists were granted U.S. Citizenship in 1954/55, and put into leadership positions within the U.S. space program due to the subversive agreement reached with the Fourth Reich:

> "... the Germans dominated the rocket program to such an extent that they held the chief and deputy slots of every major division and laboratory. And their positions at Marshall and the Kennedy Space Center at Cape Canaveral, Florida, were similar to those they had held during the war," wrote Hunt. The Peenemunde team's leader, Wernher von Braun, became the first director of the Marshall Space Center; Mittlewerk's head of production, Arthur Rudolph, was named project director of the Saturn V rocket program; Peenemunde's V-2 flight

test director, Kurt Debus, was the first director of the Kennedy Space Center."[208]

U.S. Military-Industrial Complex works with Antarctic based Fourth Reich

The role of the Dulles brothers in the secret negotiations that culminated in the emergence of the Fourth Reich during the Eisenhower administration cannot be underestimated. As described earlier, both John Foster and Allen Dulles had much experience working with German industrialists whose companies they represented when they were employed at Sullivan and Cromwell. The Dulles brothers played an important role in Hitler's ascent to power by advocating the interests of German industrialists who were united in their opposition to Communism and advocacy of a strong nationalist leader.

As the head of the Office of Strategic Services mission in Bern, Switzerland during World War II, Allen Dulles was the key U.S. official who negotiated with senior Nazi leaders to reach the secret deals that laid the foundations for later negotiations with the Antarctic based Germans. Since the Antarctic colony was led by the Thule Society and other German Secret Societies that had infiltrated Hitler's Third Reich, they were responsible for manipulating Hitler into providing the materials and personnel necessary for building the bases there. The German companies that were owned or controlled by the aristocratic members of the Thule and other German Secret Societies were well known to the Dulles brothers. Members included the Thyssens, Krupps, Flicks, Siemens, and others who owned the German companies whose subsidiaries were building spacecraft for the Fourth Reich in Antarctica.

As Secretary of State (1953-1959), John Foster Dulles was a key figure in handling the diplomatic aspects of the secret negotiations, and reaching a final agreement. His brother, Allen Dulles, as CIA Director (1953-1961), was responsible for handling all the covert aspects of the negotiations, especially when it came to implementing key details of those agreements. This included facilitating Operation Paperclip and the movement of German scientists into the U.S. who would act as intermediaries between German and U.S. companies.

During the secret negotiations, the Dulles brothers were ably supported in the U.S. Congress by Senator Prescott Bush (1952-1963), who as described in chapter two, also had extensive experience in working with German companies on behalf of prominent U.S. banking interests prior to and during World War II. Bush was also among those dealmakers who facilitated the movement of funds and resources between German and U.S. companies. Therefore, an extensive network of U.S. bankers and industrialists worked with their German peers in ways that made a mockery of the 1917 Trading with the Enemy Act that had been passed by Congress.[209]

The Dulles brothers, Bush and many other U.S. bankers, industrialists and public officials made it possible for a confluence of U.S. and German companies to reach agreements that would establish the Fourth Reich and the U.S. military-industrial complex as partners. Within a strictly compartmentalized security system, large U.S. companies worked with their smaller but more senior "German partners" in expanding out the industrial manufacturing facilities in Antarctica. However, the U.S. companies were often left in the dark about the full details of how their products would be used. The final result of this system was that fleets of antigravity spacecraft were built in Antarctica under full German control; not only for interplanetary colonization, but also interstellar conquest alongside their Reptilian/Draconian partners. Both Tompkins and Goode have described the threat this "Dark Fleet" posed for different human-looking extraterrestrial civilizations.[210]

The US Air Force and US Navy took very different approaches

in playing catch up with the Fourth Reich's secret space program out of Antarctica. Hoping to learn how to reverse engineer alien technologies, the US Air Force chose to work closely with the Fourth Reich and their extraterrestrial allies. Whistleblowers such as Charles Hall, Bill Uhouse and David Adair have revealed different aspects of the Air Force collaboration with extraterrestrials and/or German Paperclip scientists in their reverse engineering efforts.[211]

Operation Paperclip scientists worked closely with the Air Force in developing rocket technologies for a future secret space program that would specialize in global surveillance and protection. Today the US Air Force, according to Corey Goode, maintains at least two stealth space stations located approximately 500 miles above Earth.[212] The Air Force uses fleets of TR-3B antigravity vehicles operating out of Area 51 to service their space stations and to act as intercept vehicles for extraterrestrial intruders.

In contrast, the US Navy decided to develop a close working relationship with human-looking or "Nordic" extraterrestrials, according to William Tompkins. These Nordics were the chief rivals of the Reptilian/Draconian extraterrestrials who helped the Nazis and German secret societies develop their bases in Antarctica. Beginning in 1942, the Navy began to work intently with the Douglas Aircraft Company in understanding the extraterrestrial spacecraft retrieved from the Los Angeles Air Raid.[213]

After October 1948, an internal division took place within Douglas, separating their in-house think tank for studying extraterrestrial technologies, called Advance Design, from another similarly tasked working group within their facility known as Project RAND. As a result, Project RAND moved to a new location to become the RAND Corporation. Corroboration that Project RAND (and by direct association, Advance Design) studied retrieved extraterrestrial vehicles is evidenced by a Majestic Document called the "White Hot Report". This key document lists

Project RAND among the research organizations studying artifacts recovered from the 1947 Roswell crash of extraterrestrial vehicles:

> Based on all available evidence collected from recovered exhibits currently under study by AMC, AFSWP, NEPA, ABC, NACA, JRDB, **RAND,** USAAF, SAG and MIT, are deemed extraterrestrial in nature. [emphasis added][214]

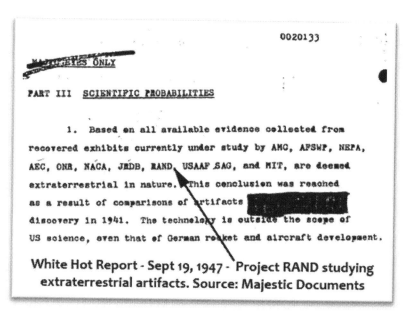

White Hot Report - Sept 19, 1947 - Project RAND studying extraterrestrial artifacts. Source: Majestic Documents

Figure 25. White Hot Report

Significantly, the leaked "White Hot Report" received the highest level of authenticity from the independent investigation of veteran document researchers, Dr. Robert Wood and Ryan Wood.[215] Consequently, the "White Hot Report" document is independent confirmation that the Douglas Aviation Company, through Project RAND, was involved in the study of retrieved alien spacecraft, just as Tompkins claimed. After the 1948 division at

Douglas, the Navy would continue to work with Advance Design, while the Air Force chose to contract with the newly created RAND Corporation. Furthermore, the Navy was aware that at least three Nordic extraterrestrials had infiltrated Douglas and were assisting Tompkins and other engineers in designing space battle groups. Both Tompkins and Goode state these battle groups were eventually deployed in the early 1980's.

Antarctica Opens Up

The secret agreement reached at the Holloman AFB meeting led to full cooperation between the Eisenhower administration and the German breakaway group in Antarctica. As a direct result, the international community was allowed to establish facilities in Antarctica, but this would be done under certain conditions designed to ensure that the German colony there would not be threatened in any way. This international cooperation was launched under the auspices of planning for the upcoming International Geophysical Year of 1957-58.

The *Antarctic Sun* describes how the international community met in 1955 to discuss development within Antarctica, especially when it came to the geographic and geomagnetic South Poles which were coveted by the major geopolitical rivals, the U.S. and USSR:

> An international conference in 1955 near the summit of Montparnasse in Paris, France, set in motion a sequence of actions that determined the scientific and political fate of an entire continent. The meeting assembled scientists from 11 nations planning IGY research in Antarctica, and their job was to decide where each nation would place its research facilities.

A dramatic event was selection of the nation to build and maintain a research station at the geographic South Pole. Vladimir Beloussov of the Soviet Union created a sensation by saying his country would place a station near the Pole. Laurence M. Gould, the U.S. delegate, had made it known the United States had a similar intention.

Sensing an opportunity to retain the prestigious location for Western researchers, the French chair, Georges Laclavère, pointed out a vast cavity in IGY coverage in East Antarctica. Behind-the-scenes maneuvers are said to have occurred. When Beloussov next took the floor he said, "We do not insist on the geographic pole." The Soviet Union took responsibility for the geomagnetic pole in East Antarctica; Vostok Station remains active there today. The United States committed itself to the geographic South Pole.[216]

The Antarctic Sun went on to explain the dramatic change in terms of numbers of personnel and stations that would be allowed in Antarctica:

The work planned for Antarctica was unprecedented. Before the IGY less than half of Antarctica had even been seen, and as late as 1955 only 179 people wintered at 20 small coastal stations operated by four nations. For the IGY, 912 people would winter at 48 stations of 11 nations, and the summer population would reach 5,000 – more than today. [217]

Another article in *The Antarctic Sun* describing the history of Antarctica stated:

In just two years the U.S. Navy built seven Antarctic stations for the IGY – five along the coast and two inland. While ten additional nations established 40 Antarctic stations of their own in the same period, between 1957 and 1958.[218]

While countries pushing into Antarctica led to the global public believing the icy continent was being developed in order for permanent bases to conduct scientific exploration and research, these programs were only a clever cover used to hide the truth. Far below the frozen ice sheets of Antarctica, there were German-Extraterrestrial bases actively collaborating with the U.S. military-industrial complex in building advanced space fleets for global dominance and interplanetary conquest. The most staggering aspect of this collaboration was that the German-Reptilian-U.S. alliance conducted horrific experiments using slave laborers, just as the Nazis had during World War II.

CHAPTER 6

Slave Labor in Antarctica

Historic use of Slave Labor in Nazi Weapons Projects

The policy of slave labor used by Nazi Germany during World War II was continued by the German companies that established subsidiaries in Antarctica to build the secret space program. Historically, throughout the war, major German companies were required to use slave labor in the industrial manufacturing process.[219] Slave labor usage was deemed absolutely essential for major German armaments companies in order to meet their war production goals. This was because of the acute labor shortage caused by widespread military conscription, by which all men of a war-fighting age had to serve on one of the multiple military fronts.

Most importantly, unlike the allied powers, Hitler's Nazi ideology required that German women stay at home "to avoid moral corruption" and to raise large families, which would also eventually provide fresh recruits for the Third Reich's armies. Hitler's powerful Armaments Minister, Albert Speer, was critical of this misplaced "romantic ideology" limiting German women's roles and wrote:

Typical of this intertwining with a romantic ideology was Hitler's, Goering's and Sauckel's [Nazi Head of Labor Deployment] refusal to let German women work in the armaments industry during the war, something that came about as a matter of course in the Anglo-Saxon countries. The reason given was that factory work would damage their morals and their child-bearing capacity. Such unsophisticated feelings were not consistent with Hitler's plans to make Germany the most powerful nation on Earth.[220]

In order to fill the growing national labor shortage due to the exclusion of war fighting German men and family bound women, foreign workers were at first encouraged to move to Germany. However, as the war progressed, the labor shortage grew and Germany continued to suffer significant manpower losses:

During the blitzkriegs of 1939/40 Germany had had no difficulties providing a sufficient number of soldiers as well as workers for its domestic economy. However, with the beginning of the invasion of Russia in June 1941, this was no longer possible. The wide fronts and the losses of the war led to the draft of more and more workers to the military. That way, 7.5 million vacancies were available despite (predominantly still voluntary) recruitment of laborers from other countries.[221]

In March 1942, Fritz Sauckel became Nazi Germany's General Plenipotentiary for Labor Deployment and immediately began "Sauckel campaigns" to bring in foreign workers using whatever means deemed necessary:

As early as 1942, approximately 2.7 million people were brought into the Reich by means of large-scale "Sauckel campaigns". Here, due to a special order by Hitler, international law was not to be considered, especially in Poland and the Soviet Union.[222]

With the growth in both prisoners of war and concentration camps, people held in these encampments were forced to work as slave laborers in German industries.

After the definite turn of the war in the winter of 1942/43, however, Sauckel was confronted with Speer's constant demands for more laborers. The number of forced laborers, often abducted with brutal means mainly from Eastern Europe and partly living under disastrous circumstances, rose up to five million. Eventually, around 20 percent of all jobs were filled by foreigners; including prisoners of war and concentration camp prisoners they amounted to more than one third. Although the demand could never be met, forced labor prevented an early collapse of the German war industry.[223]

German companies increasingly used slave labor to fulfill production targets, which were essential for their survival in wartime Germany. Company executives who failed to fulfill designated war production quotas risked military conscription and dreaded reassignment to the Russian front.

Major companies such as I.G. Farben, Siemens, Volkswagen, BMW all used slave labor. Decades after World War II, these companies or their successors agreed upon a compensation fund for former slave laborers. The following is a summary of the major companies involved in the compensation fund which was announced in February 1999 by the German government:

> After the 1998 elections, the newly elected government ... pledged to set up foundations to handle financial compensation. Twelve German industrial giants (*Allianz, BASF, Bayer, BMW, Daimler Chrysler, Degussa-Huels, Dresdner Bank, Fred Krupp, Hoesch Krupp, Hoechst, Siemens,* and *Volkswagen*) met with German Chancellor Gerhard Schroeder in [Feb] 1999.... They later announced the establishment of a fund to pay their victims. News reports speculate that it might have amounted to about 3 billion German Marks, or 2.6 billion U.S. dollars. Chancellor Schroeder saw the fund as a win-win situation for both surviving Nazi victims and German industry. He said: "*for those [victims] in the final years of their lives, it will ... provide them with a little more means that they would otherwise have had.*" German industry will probably save money because the companies would expect to be given immunity from future class-action lawsuits. Paying into a multi-billion dollar fund is probably cheaper than meeting any future financial awards by the courts.[224]

German companies fulfilled orders not just for Speer's Ministry of Armaments, but also for Himmler's Nazi SS. After his release from Spandau Prison in 1966, Speer wrote a book about the parallel industrial empire that was developed by Himmler's SS. In *Infiltration: How Heinrich Himmler Schemed to Build an SS Industrial Empire,* Speer describes how he was ordered to support Himmler's SS in building its parallel industrial empire that used millions of slaves for building super weapons in huge underground construction facilities:

Some 14.6 million slaves working to carry out Hitler's and Himmler's construction plans: a human lifetime later, this seems like a sheer pipe dream. But we must not forget that between 1942 and 1945, Sauckel managed to deport 7,652,000 people from the occupied territories to Germany in order to use them for German industry.[225]

When it came to top secret weapons such as the V-2 [aka A-4] rockets that were initially under the authority of Speer, his preference was to use German laborers in order to ensure secrecy and prevent foreign espionage. He explains in *Infiltration*:

> On July 25, 1943, Hitler signed an edict prepared by myself: "The greatest output of A-4 missiles is to be attained as swiftly as possible.... The German plants that manufacture the A-4 missile, as well as those supply the component parts, must instantly be supplied with skilled German workers ... The Reich Minister for Armaments and Munitions [Speer] is to direct the A-4 program." My authority ... was unrestrained. Only German workers – as Hitler had first determined two months earlier – were to implement the project; and we were to avoid Sauckel's program of forced laborers, which would simply promote the infiltration of spies.[226]

However, Speer's authority over the V-2 rocket program lasted only four weeks due to the destruction of the Peenemunde facility where the rockets were built. The following describes the August 17/18, 1943 bombing raid that destroyed much of this top secret facility:

> The first Crossbow target hit was Peenemunde. The primary objective of the raid was to kill as many

personnel involved in the V-weapons programs as possible, so the housing area was the main aim point. Two lesser objectives were to destroy as much of the V-weapons related work and documentation as possible, and to render Peenemunde useless as a research facility. On the evening of 17/18 August 1943, with the backdrop of a full moon, Bomber Command launched 596 aircraft – 324 Lancasters, 218 Halifaxes, 54 Stirlings – which dropped nearly 1,800 tons of bombs on Peenemunde; 85 per cent of this tonnage was high-explosive.[227]

Speer's secrecy system for the V-2 and advanced rocket program at Peenemunde had clearly failed, and responsibility for the program was quickly reassigned to the Nazi SS. Himmler had diametrically opposed ideas on the efficacy of using forced labor to ensure security for top secret research projects. Himmler's ideas impressed Hitler, as Speer notes:

> At some point Hitler brought up the manufacture of the A-4, and the necessity of keeping it top secret. In this respect, he said he had received a cogent suggestion from Himmler: Our concern about any betrayal of this highly crucial armaments project could be reduced to a fraction if the work were done by prisoners of concentration camps. Himmler, he said had told him that he could guarantee all the necessary manpower for the project. Skilled workers and even scientific specialist would be removed from the concentration camps and used in the construction of the rockets. Furthermore, he had asked a young, energetic construction expert, who had already proved his outstanding ability, to take charge of the

enterprise.[228]

Thus, the dye was now cast for future top secret projects run by the Nazi SS – forced labor was the desired choice since it was expendable. Forced laborers were closely guarded in adjacent concentration camps, and at the end of a research project, they were either transferred to a new project or put to death.

Advanced Weapons Projects using Slave Labor go Underground

The young "energetic construction expert" who took charge of the V-rocket program was Dr. Hans Kammler. Through talent and aggressive cunning, he rose quickly to become Speer's nemesis, and Himmler's desired replacement for Speer as Armaments Minister. Speer writes:

> Himmler succeeded with his decisive stroke against my previously unchallenged authority by the circuitous route of developing and manufacturing the new rocket. Kammler, who made an extremely fresh, energetic, and ruthless impression ... had begun by taking over relatively small tasks of the A-4 production within the overall armaments area. Then he had assumed responsibility for rocket launchings which was actually a military task. Finally, he obtained the production of all special weapons on the basis of rockets, and at the close of the war he had also received responsibility for manufacturing all jet airplanes. At the last minute, Hitler put Kammler in charge of all air armaments. Thus – just a few weeks before the end of the war – he had become commissioner general for all

> important weapons. Himmler's goal was achieved. But there was no more armaments industry.[229]

Speer's admission here is very significant since he verifies that he was taken out of the loop when it came to advanced weapons projects which were all placed under the control of Kammler and the Nazi SS. In addition to V-Rockets, Kammler and the SS built flying saucer prototypes for the war effort in highly restricted underground facilities. It was in the construction of these massive underground facilities that Kammler excelled, and thereby, was able to rise extremely rapidly in the Nazi hierarchy, where he eventually eclipsed Speer's authority as Armaments Minister.

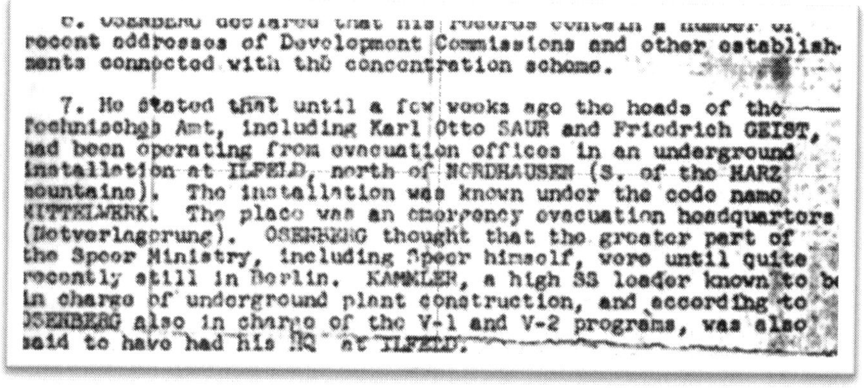

6. OSENBERG declared that his records contain a number of recent addresses of Development Commissions and other establishments connected with the concentration scheme.

7. He stated that until a few weeks ago the heads of the Technische Amt, including Karl Otto SAUR and Friedrich GEIST, had been operating from evacuation offices in an underground installation at ILFELD, north of NORDHAUSEN (S. of the HARZ mountains). The installation was known under the code name MITTELWERK. The place was an emergency evacuation headquarters (Notverlagerung). OSENBERG thought that the greater part of the Speer Ministry, including Speer himself, were until quite recently still in Berlin. KAMMLER, a high SS leader known to be in charge of underground plant construction, and according to OSENBERG also in charge of the V-1 and V-2 programs, was also said to have had his HQ at ILFELD.

Figure 26. Intelligence Objectives Sub-Committee Report (number 51) June 2, 1945. Identifies Hans Kammler in charge of Underground Facilities.

Speer emphasizes Kammler's ruthless efficiency in building massive underground facilities for the V-2 rocket and various advanced technology programs. The V-2 Rocket program at the Nordhausen facility in the Harz Mountains is the most well-known of Kammler's massive underground construction projects. Here, Werner Von Braun and other top Nazi scientists worked on advanced rocket technologies, which were used later to kick-start the U.S. missile program and the NASA space program. Slave labor was extensively used at Nordhaussen, in conditions that came to resemble Dante's *Inferno*, as Speer vividly recounts:

> In the Central Works, the cave district in the Harz Mountains, where the production of the A-4 was being prepared, conditions were scandalous and actually interfered with production. In early December 1944, Dr. A. Poschmann, chief physician of the Todt Organization, told me that he had seen Dante's Inferno. A few days later ... I went to inspect the production personally... What did I see? Expressionless faces, dull eyes, in which not even hatred was discernible, exhausted bodies in dirty blue-gray trousers.... The prisoners were undernourished and overtired, the air in the cave was cool, damp, and stale and stank of excrement. The lack of oxygen made me dizzy; I felt numb.[230]

Figure 27. Slave Labor at Nordhaussen

Despite the atrocious slave labor conditions, Speer was nevertheless tremendously impressed by Kammler's ability to turn large cavern systems into efficient production facilities, while overcoming the many inherent obstacles in utilizing the caves effectively for manufacturing purposes:

> Kammler, who was an engineer with a university degree, displayed abilities where the SS bureaucracy of the concentration camps had failed. His success was sensational in regard to A-4 manufacturing. "In an almost impossibly short period of two months, he [transformed] the underground facilities [in the Harz Mountains] from a raw state into a factory" I wrote him that this feat "does not have an even remotely similar example anywhere in Europe and is unsurpassable even by American standards".[231]

Soon after the decisive defeat at the battle of Stalingrad (August 23, 1942 to February 2, 1943), the decision was made to build concrete bunkers for the Nazi leadership, and to move heavy industry into underground facilities where both would be protected from allied bombing campaigns which were rapidly growing in intensity. Speer explains this decision and Himmler's decisive role in initiating the underground projects and making them a part of a grossly ambitious vision:

> Hitler had agreed with Himmler over a year before that [March 1943] about the necessity of the concrete protection. Now he [Himmler] only repeated his old demand when he told Dorsch: "The measures initiated" for the safety of the armaments industry by means of caves and bunker construction "are not to be carried out as temporary measures under any circumstances....

They are the prelude to a far-reaching and definitive transfer of all German industrial factories under the earth, since this is the only way to create long-term conditions for preserving the manufacturing plants in case of war.[232]

Himmler ordered an extensive study of Germany's cave system to determine the feasibility of a vast underground network of industrial projects, many of which would be top secret advanced weapons projects under the direct control of the Nazi SS. The cave research project was placed under the authority of the Nazi Ahnenerbe, which was filled by Thule Society members who had initially funded the Vril Society's flying saucer projects:

> In August 1943, he [Himmler] turned to ... the "Cave Demonstration Division in the Military Science Institute for Karst and Cave Research of the SS Karst Defense Unit." Himmler ordered this bureau to draw up a register of German caves. The 11-page report was divided according to states. It listed 93 caves ... A few weeks later the SS Ancestral Heritage Office [Ahnenerbe] was told to "confer with all the existing experts on the homeland as well as other private scholars, of whom there must be a considerable number," and they were to work out a scientific compilation of the existing caves.[233]

Himmler ultimately placed authority for the construction of the underground facilities for SS armaments and "super weapons", which included flying saucer research, with Kammler. Even Speer, in grudging admiration of Kammler's underground construction accomplishments, entrusted him to build similar facilities for the German armaments industry more generally:

Kammler's astonishing performance induced me to entrust him – as I informed Himmler on December 22, 1943 – "with special construction assignments…. In early March 1944, Goering too, still in charge of air armament, appointed Kammler his representative for the "Kammler special constructions".[234]

What is instructive here is that the same skills used to build massive underground manufacturing facilities in Nazi Germany would be essential for creating similar facilities in Antarctica. It is therefore not accidental that Kammler was among the Nazi SS officials who escaped from Nazi Germany as Hitler and the Nazis moved all resources to South America and Antarctica.

Antarctica & the Post-war Nazi Plan for a Slave Economy

Hitler authorized the Nazi SS to prepare plans for a post-World War II economy which would continue to use slave labor in an era of peace between the two remaining military powers; Nazi Germany and the United States. With a hypothesized Nazi victory over the Soviet Union, and an end of military hostilities in Europe, the Third Reich would begin to rebuild the European economy while also preparing for a future conflict with the United States.

Not only had the SS used millions of slaves for the war effort, but they planned to continue to use slaves in peacetime to prepare for the inevitable military confrontation with the U.S. Speer describes the Nazi SS vision of a post-war slave labor economy:

This vision of a "peacetime Reich" was thus based on the existence of millions of permanent slaves, who were neither political opponents nor so-called "racial enemies." Because of economic necessity, they would be kept in camps all their lives – with "women in brothels." This empire of slaves, which was to stretch all the way to the Urals, would be the basic energy source of a Europe that had to prepare to conquer the greatest enemy: the United States of America.[235]

In his book, Speer emphasizes the importance of slave labor as a key aspect of Himmler's planned future. Slave labor practices would continue even while Nazi Germany sought to establish political and economic hegemony in a post-World War II competition with the United States, which could likely last for decades.

To date, Speer's book has not been widely read or discussed since it only seemingly deals with a global political and economic situation which did not arise due to Germany's apparent military defeat. However, *Infiltration* offers a first hand account of events confirming how the Nazi economic resources were secretly distributed around the world making the emergence of the Fourth Reich as a covert global military and economic power possible, while concurrently establishing a secret space program in Antarctica. The Fourth Reich's power would be exercised covertly, and in its secretly constructed bases of operation, slave labor would continue to be used.

Crucially, Speer's book accurately describes the political economy that was used by the German Antarctic colony in the post-World War II era. As German operations in Antarctica expanded into space, slave labor continued to be exploited in their quiet preparations for a looming future confrontation with the U.S. It is very possible that Speer wrote *Infiltration* as a covert warning because he understood the coming grave world consequences of the plans for a Fourth Reich based on super weapons developed by the German Antarctic program.

William Tompkins, in a private interview, offered further

details of how slave labor was transported down to Antarctica:

> The Germans were already developing the capability to operate out in space. Go into the moon. Go to Mars and with the plans to go out into the Galaxy. And so, what took place was that everybody that was on the program, whether it was in Maria's [Orsic] side or whether it was the SS, the information eventually gets all to the German side, not hers, and gets into manufacturing. So, if you can visualize a dozen different classes of naval-type spaceships being built in production facilities in ground, not underground, but in ground mountains all over Germany and then in the occupied countries, massive numbers of these different extraterrestrial vehicles reengineered and modified to our capability or their capabilities to building it, and then learning how to fly to join the Reptilian navy's mission out in the galaxy. So now, you've got thousands of machinists and thousands of production people building different parts of these extraterrestrial vehicles.
>
> And these are inside of mountains, in deep production facilities like aerospace companies. And these people were all stopped from working. They were put on to large submarines, all of these workers which were slaves, and taken to Antarctica. And along with them, every drilling machine, every blade, every saw, every kind of manufacturing equipment necessary to build these things was picked up in one stop and taken out of Germany or out of the other countries and taken to Antarctica. So, 80 to 90 percent of this was gone a year before the war stopped.[236]

German companies that had established subsidiaries in Antarctica began producing the many components for the fleets of spacecraft essential for the ambitious space missions being conducted to the Moon, Mars, the asteroid belt and even beyond the solar system. Production of the various Haunebu, Vril and Andromeda series of spacecraft increasingly accelerated,[237] while German scientists continued to make major breakthroughs in propulsion, navigation and material sciences.[238]

Among the many thousands of UFO sightings seen around the world after the failure of Operation Highjump in early 1947, many of these were in reality spacecraft secretly operating out of Antarctica. Indeed, Admiral Richard Byrd, the leader of that fateful operation had publicly warned of an enemy that could fly from pole to pole in his 1947 interview with the Chilean press, before being muzzled from doing further media interviews by the Pentagon.[239] In fact, Byrd was referring to the German Antarctic colony which had ably shown off its achievement in weaponizing flying saucer technologies during its military engagement with his US Navy fleet.

Overflights by flying saucers upon U.S. territory and around the world have involved human-built craft based on early Nazi designs according to most authors who have researched the connection between Germany's secret flying saucer programs and the modern UFO phenomenon. These include the authors of seminal books on the topic, such as Rudolf Lusar's *German Secret Weapons of the Second World War* (1956) and Renato Vesco's *Intercept UFO* (1968).[240] More recently, Henry Stevens, author of *Hitler's Flying Saucers* (2003), and Joseph Farrell, author of *Nazi International* (2013), also came to the conclusion that most UFO sightings involve man-made craft based on early Nazi designs.[241] In this regard, Stevens writes:

> The government has used "flying saucers" to cover its own testing of secret aircraft. It uses the UFO-extraterrestrial ploy superbly. When a UFO is seen

by civilians, a controlled procedure is enacted. This procedure plants or encourages witnesses who expound an extraterrestrial origin in a given sighting.[242]

The above four authors collective view that many UFO sightings are in fact human built craft based on early Nazi flying saucer designs in secret projects is only partially accurate. Lusar, Vesco and Stevens have not considered the possibility that the Nazi's had survived the war in remote locations, while Farrell believes that this has indeed happened. His book, *Nazi International,* makes the case that the Nazis successfully relocated substantial material, wealth and manpower to South America. However, he found it a "very unlikely possibility" that a German base in Antarctica had been built.[243]

What the preceding four authors did not consider or accept, however, is that the Germans secretly built bases in Antarctica from which their flying saucer developments continue to the present day. The major exception has been W.A. Harbinson, who wrote a fictionalized account of the Nazis surviving World War II in an Antarctic fortress. In his introduction to the 1995 edition of Renato Vesco's book, he wrote about the chief premise of his novel, *GENESIS*:

> Regarding the possibility of the Germans building self-sufficient underground research factories in the Antarctic, it has only to be pointed out that the underground research centers of Nazi Germany were gigantic feats of construction, containing wind tunnels, machine shops, assembly planets launching pads, supply dumps and accommodation for all who worked there, including adjoining camps for the slaves - and yet very few people knew that they existed.
>
> Given all this, it is in my estimation quite

possible that the men and materials were shipped to the Antarctic throughout the war, that throughout those same years the Germans were engaged in building enormous underground complexes in *Neuschwabenland* similar to those scattered around the last redoubt, and that the American, Russian and British "cover-up" regarding saucer sightings could be due to the reasons given in this novel.[244]

All the preceding authors (Lusar, Vesco, Stevens, Farrell and Harbinson) emphasize the terrestrial nature of the German flying saucer programs, and the subsequent UFO phenomenon beginning in the post-war era. Importantly, what they did not consider was the extent to which the Germans had been helped by different extraterrestrial races in the design and development of flying saucer craft.

Many German scientific breakthroughs were due to the assistance they received from Reptilian extraterrestrials, according to William Tompkins. He says that US Navy spies reporting to Naval Air Station, San Diego described agreements reached between Reptilian extraterrestrials and Hitler:

> The US Naval operatives in Germany found out about the Reptilian extraterrestrials advising Hitler and the SS in Germany as to how to build massive spacecraft carriers and space cruisers to operate with the Draco Reptilian space navy. Now, what this turned out to be, after I got deeply into this, was this is the first time anybody in the United States knew that Reptilian extraterrestrials were actually here on the planet, and were working with Hitler and the SS in Germany, and had signed legal agreements of cooperation with them. [245]

Chapter four discussed Tompkins' claim that the Reptilians had assisted the Nazi regime in locating suitable locations for future bases in Antarctica, where they began to build fleets of antigravity spacecraft. Here, he offers further elaboration on the extensive underground facilities built in Antarctica:

> By the end of the war, Germany had already built massive under-mountain research facilities, not underground, but under-mountain. The word 'underground' shouldn't be used to describe this. All over Germany, and in the occupied countries, Germany had been building mass production facilities for a dozen different types of extraterrestrial vehicles that were given to them by the Draco Reptilians... So these designs were going into mass production, not just prototypes. They were going into mass production using slave labor with many massive in-mountain facilities. Eighty percent of these facilities had all been removed from Germany six months before the war ended. It had all been taken to caverns in Antarctica, and the Germans were continuing the construction down there.[246]

Germans with verifiable Aryan ancestry were selected as the senior officers, crew, scientists and engineers for the spacecraft secretly developed and flown out of Antarctica. The key element of the Thule and Nazi ideology of a master race destined to rule humanity continued to be a cornerstone of belief for the Antarctic based Germans. This meant that those with inferior genetic stock would at best be assigned to provide support services in terms of junior positions such as crew or technicians, and at worst, as slave labor. This was made possible due to the way in which Antarctica Germans rationalized the defeat suffered by Hitler's Third Reich in World War II.

Hitler had led Germany into disaster due to his reckless militaristic policies which culminated in the beginning of World War II. This was well before German scientists and corporations had mastered the advanced technologies that were then being developed, and deemed necessary for a successful military campaign. It is well known to historians that leading German figures such as Fritz Thyssen, head of Germany's major steel conglomerate, and Admiral Wilhelm Canaris, head of Germany's military intelligence, were opposed to Hitler's militarism.[247]

Thyssen, Canaris and other members of Germany's secret societies, who had elevated Hitler to his leadership position in 1933, believed Germany needed more time to develop its industrial might through the advanced technologies being researched and developed by German companies. If Hitler had not pursued his reckless militaristic policies, Germany's Third Reich would have had more time to develop its advanced technology programs, which could have possibly changed the outcome of World War II.

The chief lesson the German Antarctic colony gained, therefore, was that in any future struggle with the United States, covert economic means needed to be used to establish global hegemony. An open military struggle was likely to have similar outcomes mirroring the two world wars, since Germany simply lacked the manpower and industrial manufacturing resources to defeat other major nations united in a clear opposition. A covert struggle, where major powers were picked off one by one, would be an easier way to establish global dominance.

A covert global struggle where German elites would collaborate with elites from other nations to expand Germany's secret Antarctic operations would have had a very different outcome. After all, it was shown in an earlier chapter how American elites such as Henry Ford proposed some of the racial theories later adopted by Hitler. Nationalism would be used by elites to manipulate the masses to eagerly fight in wars, but for the elite groups themselves, "the Fraternity", what motivated them was not nationalism, but the racial supremacy views they all shared in their unending quest for personal power.

It was not just German secret societies that believed in Edward Bulwer Lytton's theories about the Vril force, futuristic societies in the Inner Earth, and advanced technologies that would in future enable space travel and global dominance. These views were shared by American, British, French and Russian elites, and similar groups from the Far East. All believed that power should be held by those born and bred to wield it, not exercised through the ballot box as widely accepted in democratic societies.

This made it possible for the German Antarctic colony to find influential allies in the military industrial complexes of major nations, which then led to the infiltration of these nations. Both U.S. elites and the Antarctica Germans perceived many benefits from cooperation since each lacked something the other side possessed. The Germans held incredibly advanced space technologies, either developed by their secret societies or given to them by their Reptilian allies, but lacked manpower and the industrial resources to fully develop these on their own. On the other hand, U.S. elites had vast manpower and industrial resources, but lacked the scientific understanding for studying and reverse engineering the extraterrestrial technologies that had come into their possession.

It is important to keep in mind that there were many U.S. companies that had established close relations with their German partners in the lead up to, and during, World War II. Influential U.S. figures such as Henry Ford, the Rockefeller family, and the Dulles brothers were the tip of the American iceberg when it came to extensive economic cooperation with Nazi Germany. After World War II, some of the more notable U.S. elites had advanced to prominent political positions that would facilitate close cooperation between the U.S. and German Antarctica colonies. In early 1953, Nelson Rockefeller had become a senior advisor to President Eisenhower and reorganized the U.S. bureaucracy. John Foster Dulles became Eisenhower's Secretary of State, while Allen Dulles became the CIA Director. These three officials were at the center of what would become a very close relationship between the U.S. and Antarctic based Germans.

In this relationship, elite U.S. policy makers would turn a blind eye to the extensive use of slave labor in Antarctica. In fact, many accepted the advantages a captive pool of humans could provide for corporations conducting research and development into bio-warfare, genetic engineering and space travel, all of which could take place without any kind of oversight. In chapters yet to come, a deeper explanation will be provided to show how slave labor would be used for Research and Development programs in Antarctica, and subsequently taken off planet in space colonies established by the Germans.

It would be a mistake to consider the German Space Program in Antarctica as exclusively militaristic, exploitative of slave labor, and intent on galactic conquest. This is because of the Orsic/Vril Society faction that also operated both in Antarctica and in South America from smaller German bases. Orsic's antipathy towards militarism and the Nazi SS led to her far smaller space program which took a very different approach in interacting with the rest of humanity. She would attempt to disseminate a galactic philosophy of peace and unity, and help raise human consciousness. In this effort, she was helped by like-minded human looking extraterrestrials; the Nordics. Together, Orsic and the Nordics played a significant role in launching the Space Brothers movement in the 1950's.

CHAPTER 7

German Secret Space Program
& the Space Brothers

Only a few months after Admiral Richard Byrd's forewarning interview with the Chilean press, the famous Kenneth Arnold UFO incident occurred. In June 1947, Arnold witnessed fleets of flying wing-shaped craft over the Cascade Mountains of Oregon and Washington State. These flying wing craft were remarkably similar in design to a unique aerial model the Horton Brothers had developed for Nazi Germany – one of which was relocated to the U.S. after World War II. It's feasible that successful prototypes were developed and moved to Antarctica, and were able to overfly U.S. territory by 1947.

Admiral Byrd's warning had proved prescient insofar as the Antarctic based Germans now had the capability to overfly the U.S. with impunity by June 1947. Therefore, with the subsequent rise of the UFO sightings after the Arnold incident, it can be concluded that some, if not many of these, were connected to the German Space Program out of Antarctica.

As already mentioned, the most memorable flyover within the U.S. occurred during two successive weekends in July 1952, when waves of UFOs filled the air space over Washington, DC.

Drawing of UFO sighted by Kenneth Arnold in 1947 compared with photo of Horton 229 developed by Nazi Germany in 1944.

Figure 28. Comparison: UFO drawing by Kenneth Arnold /Photo of Horton Flying Wing

Tens of thousands witnessed the flyovers which were photographed, caught on radar, and sighted by military pilots. The sightings were so dramatic that the US Air Force gave a Press Conference to dismiss it all as a "temperature inversion" weather anomaly. According to three present day insiders/whistleblowers who acquired knowledge of the mysterious crafts' origins, the UFOs were in fact German flying saucers. William Tompkins, Clark McClelland, and Corey Goode all said the same regarding the Washington Flyover. Each of them had either been briefed or had learned from colleagues that the UFO vehicles were German antigravity spacecraft.

Did German Astronauts pretend to be Extraterrestrials when meeting Contactees?

Only months after the highly dramatic 1952 Washington Flyover, the legendary contactee George Adamski began witnessing and photographing flying saucers over California skies. On November 20, Adamski claimed he traveled to a remote location near Desert Center, California, where he made contact with the occupant of a landed saucer ship. Six people that accompanied Adamski saw two UFO ships on the day of Adamski's encounter. The first was a large cigar-shaped craft that flew overhead; the second was a smaller saucer-shaped scout craft that landed. An occupant emerged to meet with Adamski, and eventually conveyed that he came from the planet Venus and was called "Orthon".

The six witnesses of the landing and Adamski's encounter with it's vehicle's occupant signed an affidavit supporting Adamski's version of events, which were subsequently published in his 1953 co-written book, *The Flying Saucers Have Landed*. One of the witnesses, George Hunt Williamson, said during a lecture:

I would like to reaffirm here that the experience, as George Adamski has related in *Flying Saucers Have Landed*, where my wife and I, along with friends of ours, were witnesses to the occurrence, happened exactly as Mr. Adamski mentions there in *Flying Saucers Have Landed*: the large craft was witnessed, and then through binoculars we did witness the other happenings about a mile away on the desert... We did see Mr. Adamski talking to someone ... at a distance. We saw the large craft. We saw flashes of light from it, from which we later learned the smaller craft had come out of the larger one. We did see a great opening in the bigger craft through which the smaller scout-ship must have originally left the larger ship.... We did see the small ship as it hovered in the saddle. [248]

Lending significant support to Adamski's claims are the photos he took of both the hovering saucer-shaped craft, and the pilot (Orthon) walking away from where the craft appeared to have landed. In 2017, the photos were digitally enhanced by Danish artist Rene Erik Olsen and first released in a book authored by French UFO researcher Michel Zirger, entitled: *We Are Here: Visitors without a Passport*. A sequence of the original photos taken by Adamski of the "scout craft" and "Orthon", together with the enhanced versions produced by Olsen, are shown here [see Figures 29 and 30].

There are a number of aspects about George Adamski's 1952 contactee case that raise the possibility that Orthon was part of a German secret space program, and/or linked to an extraterrestrial alliance which had actively helped Nazi Germany leading up to, and during, World War II.

Nov 20, 1952 Adamski Photo of Desert Center Landing

Image 1. Original frame taken by George Adamski with a craft showing just moving above a hillside

Image 2. Raw image enhancement of part of the frame with the craft visible

Image 3. Finished enhancement of the sky, landscape and craft

Image 4. A zoomed version of image 3.

(C) Rene Erik Olsen 2017

Figure 29. Scout Craft photographed by George Adamski. "Original images copyright The Adamski Foundation and enhancements copyright Rene Erik Olsen."

Figure 30. Two photo enhancements showing Orthon walking from landed Scout Craft and Orthon close up. "Original images copyright The Adamski Foundation and enhancements copyright Rene Erik Olsen."

The flying saucers that Adamski photographed very closely resembled the Haunebu antigravity craft that had allegedly been developed in Nazi Germany. The exact specifications of the Haunebu craft contained in Nazi SS files were first released by Vladimir Terziski, an engineer and former member of the Bulgarian Academy of Sciences who was discussed in chapter three. The following image [Figure 31] is a comparison of a 1943 design of a Haunebu II flying saucer craft being developed by the Nazi SS for the war effort, and a scout craft photographed by Adamski in December 1952.

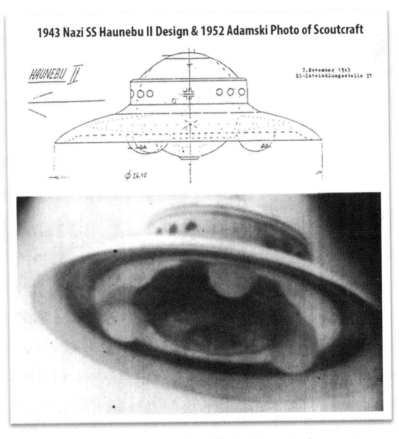

Figure 31. Haunebu II and Adamski Scout Craft

Strikingly, the only major point of difference is that the Nazi craft was equipped with an artillery device signifying an attempt to weaponize flying saucers for the war effort. Indeed, in 1950, articles appeared in major newspapers around the world citing interviews with prominent Italian and German scientists who confirmed that the Axis powers had been cooperating in a secret effort to weaponize flying saucer prototypes.[249] The configuration of the 1943 Haunebu craft is so similar to those Adamski photographed in 1952 as to suggest that they are the same type of vehicle, or whoever developed the latter was at a comparable technological level of development to the Germans in the mid-1940's.

This fact raises a big problem! If Orthon was from Venus, and part of an advanced interplanetary association, as believed by Adamski, then how could the first generation of Nazi flying saucers have been almost identical to craft used by a far more technologically evolved interplanetary society? One explanation is that extraterrestrials handed over their own craft to Nazi Germany so that Third Reich scientists could reverse engineer them. This is supported by information provided by William Tompkins, who says that US Navy spies had reported during their debriefings from 1942 to 1946 that the Germans had been given over a dozen models of operational antigravity spacecraft.[250] The Nazis were furiously attempting to reverse engineer these for the war effort, but were ultimately unsuccessful as far as their European based programs were concerned. Tompkins' information helps us to understand that the 1943 Haunebu II diagram was an attempt by the Nazis to weaponize an antigravity flying saucer, which had been given to Hitler's Third Reich by extraterrestrial allies as a result of a secret agreement. Yet, how can the lack of weapons on the flying saucers sighted by Adamski in 1952 be accounted for if they were associated in some way with the German Space Program?

Since the Germans had not one, but two, space programs in Antarctica, the answer is clear. While one of these was a fusion of German Secret Societies (including Thule, Black Sun and Nazi SS) ruled by a male dominated hierarchy, the other space program was led by Maria Orsic and her Vril Society which, to the contrary, was governed by women. Indeed, Orsic was adamantly opposed to the Nazi war effort and militarizing flying saucer technologies, as attempted by the Nazi SS. Orsic's lack of assistance towards the war effort was noted by Himmler, as pointed out in chapter one, but his protests were rebuffed by Hitler because he recognized her knowledge and extraterrestrial connections as highly valuable to the emerging German space programs. Tompkins commented in an interview on this subject:

> Germany found out about the blonde [Orsic], took her over, stopped everything, and then got to this point where there was some sort of pressurized program by the SS to control that original group. Now, several times they did work together, but Hitler allowed them to operate independently of the whole SS program – the whole development. So we had two developments going on in Germany. The girls didn't want their vehicles to be used for anything else but travel. They were afraid that somebody would get a hold of it and they'd use it for military, which is, of course, what they got.[251]

Orsic was deeply dedicated to learning about extraterrestrial life in the galaxy and cosmic philosophy, as evidenced in her psychic channeling of beings from the Alderbaran star system. This strongly indicates that the craft photographed by Adamski belonged to the Orsic/Vril faction of the German secret space program operating out of Antarctica.

Certainly, the cosmic peace and unity message advocated by Adamski's alleged Venusians had much in common with Orsic's pacific approach to learning about extraterrestrial life. Indeed, if Orsic had succeeded in using advanced space-time technologies to travel to Aldebaran and interact with advanced human colonies there in the late 1940's, then it's very possible that the Orsic/Vril faction of the German Space Program became spiritually elevated due to their interactions. Thus, after acquiring unprecedented galactic knowledge and experience, her faction returned to Earth to start promoting a cosmic peace and unity message. So, were the "Space Brothers" that dominated the 1950's UFO literature converted Germans who had spent decades learning cosmic secrets in deep space during their mission of advanced space-time travel?

The first person to raise such a possibility was legendary UFO researcher, Lt. Col. Wendelle Stevens, who passed away in 2010. In our private conversations during a two week tour of Japan in 2007, Stevens told me that Hitler survived World War II, had lived in Colombia, and that the Germans converted during their space encounters. He said that the Germans had come back to teach a message of cosmic peace and unity to the rest of humanity. In a similar vein, Corey Goode said he learned that the Orsic group did travel into deep space and had returned to Earth to preach a cosmic philosophy to contactees:

> Many in the Intel Community became quite convinced that Maria Orsic was one of the "Blondes" that would land in UFOs and talk to people in German pretending to be an ET from another star system. When some of the witnesses were shown her photo they identified her as the same person that they had met from the Flying Saucer. She has obviously made it to the Antarctic Bases/Cities where she was taking part in a program using the NAZI's Flying Disk technologies to spread disinformation through some contactees.[252]

However, the one condition imposed upon them in doing so was that they had to pretend to be extraterrestrials and not reveal their true origin. Therefore, it is very possible that Orthon had also only feigned to be from Venus, in order to hide the existence of the German space program that had survived World War II.

A fact that strengthens this conclusion is that Orthon only communicated with Adamski non-verbally using sign language during the 1952 Desert Center encounter. Adamski explained in a lecture that Orthon "spoke mostly in a strange dialect wholly unintelligible to Adamski".[253] Was the "strange dialect" High German which another contactee, Reinhold Schmidt, identified as the language used by the spacecraft occupants during his 1957 contact experience?

Michael Zirger, author of *They are Here*, summarizes Schmidt's encounter:

> I would like to quote briefly a last case, that of Reinhold O. Schmidt, 60 at the time, a Bakersfield (Calif.) grain buyer. On November 5, 1957, he claimed to have spoken for about 30 minutes to the crew of a large silver cigar-shaped UFO that had allegedly landed near Kearney on the Nebraska prairie to make repairs. In an available one-hour-and-a-half tape-recorded interview Schmidt stated the crew was composed of "four men and two ladies." They all spoke to him "in American language with a German accent," but at times he seemed to him that he could hear them talking among themselves in "high German language, very good high German." [254]

Schmidt had German parents and was also taught High German at school, but Adamski may not have been able to identify a German dialect since his background was Polish-American. Alternatively,

Adamski may have known that Orthon spoke German, but was not allowed to publicly reveal this for national security reasons.

It has long been rumored that after each of his extraterrestrial contact experiences, Adamski was secretly flown from California to the Pentagon by the US Air Force where he was debriefed about his contacts. In May 2009, a rare video was released online containing interviews with various witnesses who knew about Adamski and his secret debriefings at the Pentagon.[255] The witnesses confirmed that Adamski possessed a military ordnance ID card which allowed him access into the Pentagon. This military ID was witnessed by several people who had worked in various positions at the Department of Defense. Among the witnesses was William Sherwood, who had previously worked for the US Army Ordnance Department and possessed his own Ordnance pass. Sherwood saw Adamski's Ordnance pass and confirmed its authenticity.[256] Sherwood's, and others, supporting testimony gives credence to rumors that Adamski was indeed secretly briefing the Pentagon about his extraterrestrial contacts.

In 1952, the Pentagon was well aware that a German Space Program had survived World War II and began maneuvers over U.S. territory. Senior Pentagon officials wanted this to be kept secret. Consequently, it is very possible that Adamski was told not to say anything about Orthon speaking German to prevent the public from learning the truth.

Pentagon Hid German Secret Space Program Link to Alien Contact Cases

The idea that some of the UFOs sighted over U.S. soil were part of a German secret space program is strengthened by two additional UFO landing incidents during the 1950's, over the same period that Adamski said he was having contact experiences with Venusians. On January 7, 1956, Willard Wannall, a Master

Sergeant at the time with the US Army, says that he saw a flying saucer land in a secluded area of Kaimuki, near Honolulu, Hawaii. He was debriefed by US Army and US Air Force Intelligence officers at Fort Shafter where he was stationed, and wrote a 32 page report of the incident to his commanding officer. While Wannall's brief description of the incident was mentioned in UFO sightings reports at that time, his detailed report about it has never been publicly released.[257]

After retiring from the Army, he wrote a book in 1967 about the incident. In it he described how he was still under national security orders not to divulge key details about the case:

> However, it may be stated without jeopardizing the safety of my family and friends, or violating any security restrictions, that we witnessed the landing near our home of a clearly defined unconventional flying object which remained under our surveillance all of forty-five minutes prior to its departure. In addition to myself, there were six other responsible, and highly respected individuals present, who viewed the details of this sighting alternately with and without the aid of high-powered binoculars.[258]

Decades later, a reporter with the Maui UFO Report interviewed Wannall before his 2000 passing, and the public was able to learn for the first time some of the key details of the UFO landing incident:

> This time the bell shaped, silver, domed, port holed craft landed in the densely wooded hills behind Honolulu. When Sergeant Wannall approached, a hatch opened. He noted the swastika and Nazi Iron Cross on both the UFO and the uniform of the occupant. The saucer pilot spoke with a German

accent, and had a Nazi uniform on him! ... Sgt. Wannall told us that escaped Nazis, who had flying saucers, had fled to South America and secret underground bases, below the Ice in Antarctica shortly before the Nazi war surrender.[259]

Wannall's description of the craft closely matches those photographed and witnessed by Adamski in 1952, and the Haunbu II craft secretly developed by Nazi Germany.

It is now understandable why Wannall's 32 page report about the 1956 incident was never publicly released under Project Blue Book. It would have revealed that a German Secret Space Program was actively overflying and landing all over the U.S., including the Hawaiian Islands. The fact that the UFO pilot spoke with a German accent, wore a Nazi uniform, and was able to land near a U.S. military base (Fort Shafter) shows that the occupants had no fear of being fired upon. In turn, this clearly signifies that some kind of agreement had been reached with U.S. military authorities concerning German spacecraft flying over and landing on U.S. soil. Since such negotiations began under the Truman administration, with an agreement finally reached in 1955, in all likelihood, the Wannall incident involved the militarized faction of the German Space Program operating out of Antarctica.

This brings us back to the Reinhold Schmidt incident. Schmidt described meeting with the occupants of a flying saucer that landed on November 5, 1957 near Kearney, Nebraska, who spoke in English with a noticeable German accent to him, and used High German when communicating amongst themselves. In his book, *Edge of Tomorrow*, Schmidt wrote that he initially believed the six occupants were German scientists:

> I thought that perhaps it might have come from Russia, and that it was manned by a crew of German scientists getting data on the first Russian

Sputnik which had been launched about a week before.[260]

After being contacted later by one of the UFO occupants, Schmidt had further encounters and changed his mind about the craft's origins. He would now refer to the vehicles crew as extraterrestrials from Saturn. Schmidt wrote about his subsequent meetings with the alleged extraterrestrials, describing how they took him on trips to learn about ancient Earth mysteries, and imparted their cosmic philosophy of peace and unity. A telling example is illustrated in what Schmidt was told about a naturally occurring metal that could be used to build spacecraft:

> They showed me how a valuable metal could be extracted from the rocks of one of the quarries. This metal is similar to that which the Saturnians use in the construction of their Spaceships. When certain improvements in our social and economic systems have been made which will qualify us to associate with those people who have already learned how to work and live together in peace and friendship, then we of Earth will be able to use this metal in the construction of Spaceships in which we also can visit other planets. [261]

Schmidt, like Adamski, may have also been intentionally misled by the inhabitants of the craft he encountered when they declared themselves of extraterrestrial origin. Despite the obvious signs that they were part of a German Secret Space Program operating long after 1945, both contactees would not have known this highly guarded information. Alternatively, again like Adamski, Schmidt may have been pressured by national security authorities to drop any public references to the spacecraft occupants being German astronauts. This latter explanation is supported by the puzzling treatment Schmidt received by local authorities after reporting his first 1957 contact. The initial interest and friendly support, which led to overnight national media exposure, dramatically shifted into outright hostility by local authorities after the arrival of two Air Force officials.

After extensive debriefings with multiple local officials, and interviews with the local and national media, the Air Force officials stepped in and suddenly Schmidt found himself pressured by the Chief of Police to recant his public testimony. Schmidt was next mysteriously jailed for two days without charges, and then he was committed to a mental institution in Hastings, Nebraska. All of this was done without allowing him legal representation. Only after Schmidt's family and employer strongly petitioned authorities was he eventually released.

After the two USAF officials intervened, heavy pressure was placed on Schmidt to change his story, which clearly suggests that key elements of it threatened national security. In fact, his release from custody was likely a result that he agreed to change important details in his story to protect national interests. Schmidt's testimony suggests, as does the Adamski and Wannall cases, that craft belonging to a German Secret Space Program were actively overflying and landing within the U.S. Any reference to the German origin of such craft was thoroughly downplayed, and instead references to extraterrestrials from Venus, Saturn or elsewhere were emphasized.

Were Billy Meier's Plejarans part of a German Secret Space Program?

In 1975, Eduard Albert "Billy" Meier began taking photographs of flying saucers that landed near his home in Switzerland. He met with the craft's occupants who said they were from the Pleiades star system, and called themselves

"Plejarans". As in the Adamski case, there was a resemblance between Billy Meier's scout craft and the Haunebu flying saucers developed by Nazi Germany. Also, as in the Schmidt case, the alleged extraterrestrials spoke in German to Meier, and preached a cosmic philosophy to him, which he subsequently disseminated in a multivolume series comprised of his contact notes.[262]

Finally, we have the principal Plejaran emissary Meier says he met in 1975, Semjase, who he described as a beautiful young woman. Semjase taught Meier many things about life in the galaxy, and her people's cosmic philosophy. Corey Goode said in a 2016 lecture that when a photo of Maria Orsic was shown to Meier, he identified her as "Semjase":

> ... when the military found out about Meier's case, they sent people over with some photographs for him to try and identify the female being he saw. He quickly pointed out one photograph, saying, "That's her! That's her!" Apparently the photo he pointed out was of Maria Orsic, the medium from the Vril Society, who was making contact with inner-Earth groups, and who played an intimate role in the pre and post World War II German secret space program.[263]

When we combine the resemblance of Meier's scout craft with the Haunebu series of flying saucers, the fact that his principal contact, Semjase, spoke fluent German, and that Meier identified Semjase and Orsic as the same person, then a clear conclusion emerges. Billy Meier's contact case involved meetings with the Orsic/Vril faction of a German Secret Space Program that was promoting a philosophy of cosmic peace and unity.

Figure 32. Maria Orsic

Conclusion

The material examined in this chapter suggests that the George Adamski contactee case, as well as the Wannall and

Schmidt cases, were very likely real encounters by U.S. citizens who interacted with members of a German Space Program operating out of Antarctica. Both the occupants of the flying saucer craft, as well as the U.S. military, hid the German connection and encouraged belief in the extraterrestrial hypothesis to explain their origin. Additionally, the Billy Meier contact case almost certainly involved contact with the Orsic faction of the German Antarctic Program.

This is not to say that *all* alleged extraterrestrial contact cases have actually involved interactions with members of the positive Orsic/Vril faction largely behind the "Space Brothers" phenomenon. After all, the Germans had been helped by different groups of extraterrestrials who had supplied them with the initial designs for space-time technologies and vehicles capable of traveling to distant solar systems, and later, provided operational spacecraft for them to reverse engineer during World War II. Different factions of extraterrestrials helped the Germans establish their Antarctica bases, and protected them during a critical period when the Allied powers threatened to overrun them.

Despite the extraterrestrial's involvement in the German Space Program, there is a need to review the 1950's and 1960's contactee cases to determine the true origins of the alleged "Space Brothers", along with the Pentagon's efforts to suppress the truth. The U.S. national security establishment encouraged the debunking of contactees such as Adamski, Schmidt, Wannall, Meier and many others, not because they feared the public learning about extraterrestrial visitation, but because they feared the public learning the truth about the German breakaway colony in Antarctica, a faction of which was closely associated with the Space Brothers Phenomenon.

The most powerful nation states of the era – the U.S., Britain, France and the Soviet Union – did not want their citizens to learn that a powerful remnant of Nazi Germany had survived the war, or that this group's technological achievements in advanced aerospace technologies had become so dominant that the former Allied powers had nothing comparable. Leaders erringly assumed that secret agreements, infiltration of the German's Antarctic facilities and the deceptions of counter intelligence would be the tools to bridge the technological gap. In the meantime, they deliberately chose to suppress the truth about the German connection to the "Space Brothers" encounters that Adamski, Schmidt, Wannall and other contactees began having throughout the Eisenhower Administration in the 1950's.

On November 8, 1960, John F. Kennedy was elected as the next U.S. President, and inherited the weighty problem from Eisenhower of how a Presidential administration should deal with the menacing Fourth Reich operating out of Antarctica, its connection to the "Space Brothers" phenomenon, and the Fourth Reich's covert control over the U.S. military-industrial complex. Kennedy's courageous attempt to assert direct Presidential authority over covert UFO related projects and to wind back Fourth Reich influence would lead to a tragic confrontation.

CHAPTER 8

Kennedy's Confrontation
with the Fourth Reich

Three days before President John F. Kennedy's January 20, 1960 inauguration, the departing President Dwight D. Eisenhower gave his famous farewell speech:

> In the councils of government, we must guard against the acquisition of unwarranted influence, whether sought or unsought, by the military-industrial complex. The potential for the disastrous rise of misplaced power exists and will persist. We must never let the weight of this combination endanger our liberties or democratic processes. We should take nothing for granted. Only an alert and knowledgeable citizenry can compel the proper meshing of the huge industrial and military machinery of defense with our peaceful methods and goals, so that security and liberty may prosper together. [264]

Eisenhower's speech was a veiled warning about the sinister influence the Fourth Reich had achieved through its infiltration of the U.S. military-industrial complex, and the danger this posed to American liberties and the incoming Kennedy administration.

President Kennedy began his new administration with some important background knowledge about what had transpired behind the scenes during World War II, including Nazi Germany's development of flying saucer craft. In July/August 1945, Kennedy accompanied James Forrestal, then Secretary of the US Navy, on his trip to Allied occupied Germany. Forrestal wanted to recruit the young Kennedy to his personal staff since Forrestal was both friends with his father, Joseph, and was genuinely impressed by the young Kennedy's precocious intellect. Kennedy was at Forrestal's side during his tour of captured advanced Nazi weapons, and during meetings with senior Allied Generals such as General Eisenhower.

Kennedy recorded his visit in a diary that was posthumously published as *Prelude to Leadership*.[265] His dairy makes it clear that he was present during Forrestal's review of advanced technologies of interest to the Navy, some of which would be repatriated under Operation Paperclip. While Kennedy's diary doesn't refer to flying saucers, it is clear that he witnessed advanced Nazi technologies, and was briefed about what he could write about and what had to be withheld to safeguard national security.

Kennedy remained on close personnel terms with Forrestal, who became the first Secretary of Defense in September 1947. Then, in a decision that shocked the general public, President Truman sacked Forrestal less than 2 years later on March 28, 1949, and shortly after Forrestal died mysteriously in May 1949, allegedly from suicide. Kennedy suspected foul play against his mentor because of what Forrestal had told him about extraterrestrial technologies and the menacing activities of the German Antarctica colony that was steadily infiltrating the U.S. military-industrial complex.[266]

As President, Kennedy came into the necessary position to gain access to classified UFO files and technologies, and their connection to Nazi Germany. He chose to retain Allen Dulles as Director of the CIA in the new administration despite Eisenhower's warning about the danger posed by the military industrial complex. Kennedy's father Joseph, the former U.S. Ambassador to Great Britain, had been an early supporter of Nazi Germany. In fact, he had made large donations directly to the Nazi Party. On May 3, 1941, J. Edgar Hoover sent a memorandum to President Roosevelt's office, which read:

> Information has been received at this Bureau from a source that is socially prominent and known to be in touch with some of the people involved, but for whom we cannot vouch, to the effect that Joseph P. Kennedy, the former Ambassador to England, and Ben Smith the Wall Street operator, sometime in the past had a meeting with Goring in Vichy, France, and that thereafter Kennedy and Smith had donated a considerable amount of money to the German cause. They are both described as being very anti-British and pro-German.[267]

Joseph Kennedy's sympathies for Nazi Germany, connections with prominent Nazis such as Hermann Goring (leader of the Luftwaffe), and funding of the Nazi movement, together lead to a remarkable conclusion. Just as Joseph Kennedy had made a deal with the Italian Mafia to support his son's 1960 election, he also reached a similar deal with the Fourth Reich![268] This would help explain why President Kennedy decided to keep Dulles, knowing full well about his connections to the Fourth Reich, and that he was also the head of the Majestic 12 (MJ-12) Special Studies Group – the secret committee appointed to steer U.S. policy on extraterrestrial life and the German Antarctic presence.

On June 28, 1961, President Kennedy wrote a Top Secret memorandum to Dulles initiating the process of getting access to

MJ-12 files:

National Security Memorandum

To: The Director, Central Intelligence Agency

Subject: Review of MJ-12 Intelligence Operations as they relate to Cold War Psychological Warfare Plans

I would like a brief summary from you at your earliest convenience[269]

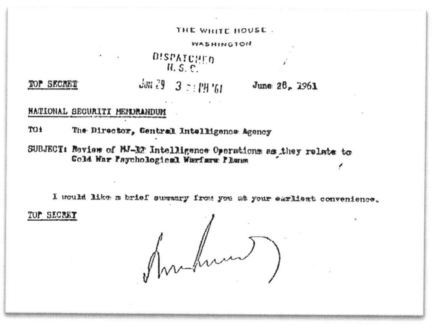

Figure 33. Kennedy's Memo to CIA Director Dulles. Source: Majestic Documents

This leaked National Security Action Memorandum clearly shows that in June of 1961, Kennedy wanted to learn about MJ-12 activities and its relationship with psychological warfare. While the leaked June Memorandum has not been acknowledged by the CIA (it was classified TOP SECRET), other documents from that period support its authenticity.[270]

Dulles' response to Kennedy's June Memorandum was an alleged Top Secret letter issued on November 5, 1961.[271] This letter gives an overview of MJ-12 activities regarding psychological warfare activities. It describes UFOs as part of "Soviet propaganda" designed "to spread distrust of the government".[272] Dulles' letter acknowledges that while it was possible some "UFO cases are of non-terrestrial origin," these did not "constitute a physical threat to national defense".[273] Most significantly, Dulles' letter says: "For reasons of security, I cannot divulge pertinent data on some of the more sensitive aspects of MJ-12 activities."[274] If genuine, Dulles' letter gave President Kennedy only superficial information in response to his June Memorandum request for a brief summary of MJ-12 activities.

Dulles & the Partially Burned MJ-12 Document

The leaked Kennedy Memorandum of June 28, 1961, and Dulles' November 1961 letter of response, reveal that a power struggle was occurring over Presidential Executive control of Cold War psychological warfare programs and the covert activities of MJ-12. Up until his resignation as CIA Director on November 29, 1961, Dulles was the pivotal figure in the power struggle with Kennedy over MJ-12 activities, including its control of classified UFO files and the rapidly growing influence of the Fourth Reich. This fight for power is reflected in a leaked draft of a memorandum, allegedly rescued from a fire burning the remainder of James Angleton's files after his death on May 12, 1987. Angleton was the CIA chief of counterintelligence (1954-1974), and was heavily involved in providing security for the MJ-12 Group and Fourth Reich activities.

In late 1974, Angleton was forced into retirement by the new CIA Director, William Colby. According to Cord Meyer, a

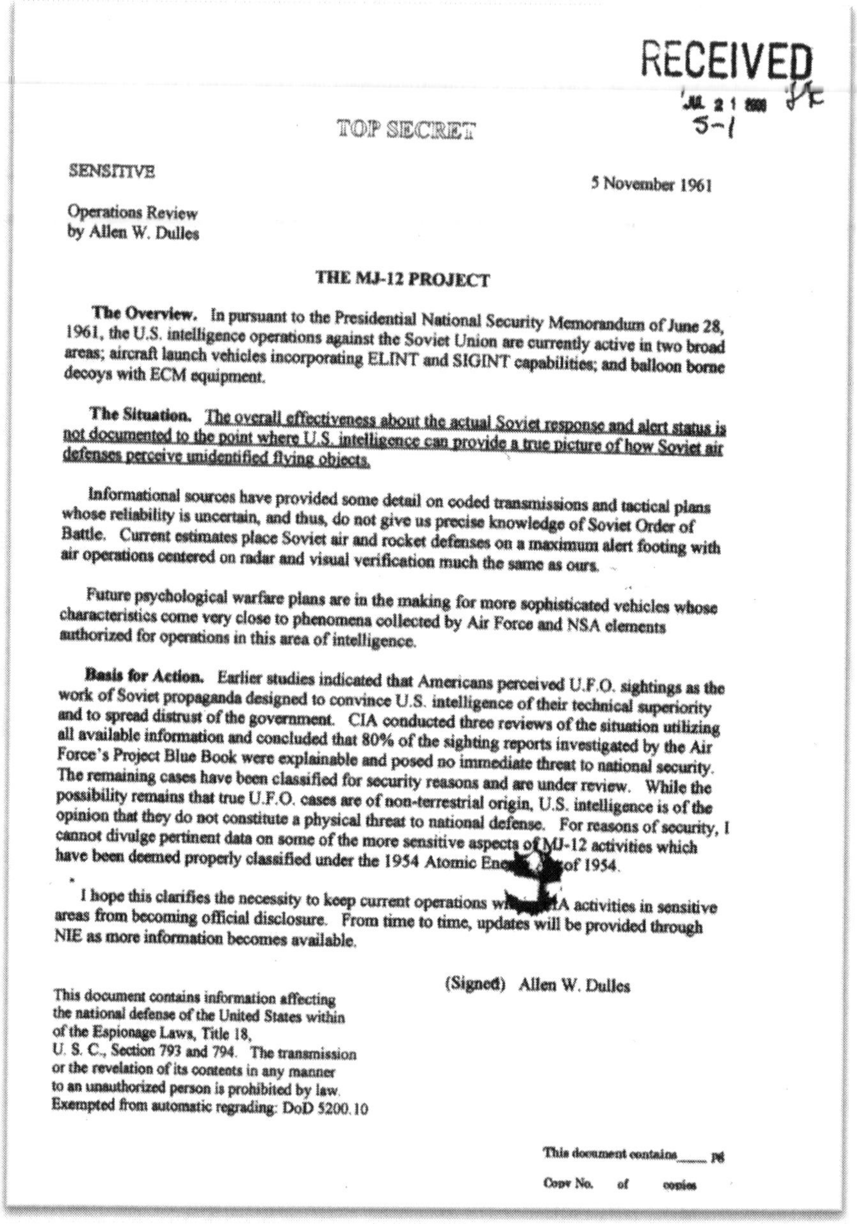

RECEIVED
JUL 21 800
5-1

TOP SECRET

5 November 1961

Operations Review
by Allen W. Dulles

THE MJ-12 PROJECT

The Overview. In pursuant to the Presidential National Security Memorandum of June 28, 1961, the U.S. intelligence operations against the Soviet Union are currently active in two broad areas; aircraft launch vehicles incorporating ELINT and SIGINT capabilities; and balloon borne decoys with ECM equipment.

The Situation. The overall effectiveness about the actual Soviet response and alert status is not documented to the point where U.S. intelligence can provide a true picture of how Soviet air defenses perceive unidentified flying objects.

Informational sources have provided some detail on coded transmissions and tactical plans whose reliability is uncertain, and thus, do not give us precise knowledge of Soviet Order of Battle. Current estimates place Soviet air and rocket defenses on a maximum alert footing with air operations centered on radar and visual verification much the same as ours.

Future psychological warfare plans are in the making for more sophisticated vehicles whose characteristics come very close to phenomena collected by Air Force and NSA elements authorized for operations in this area of intelligence.

Basis for Action. Earlier studies indicated that Americans perceived U.F.O. sightings as the work of Soviet propaganda designed to convince U.S. intelligence of their technical superiority and to spread distrust of the government. CIA conducted three reviews of the situation utilizing all available information and concluded that 80% of the sighting reports investigated by the Air Force's Project Blue Book were explainable and posed no immediate threat to national security. The remaining cases have been classified for security reasons and are under review. While the possibility remains that true U.F.O. cases are of non-terrestrial origin, U.S. intelligence is of the opinion that they do not constitute a physical threat to national defense. For reasons of security, I cannot divulge pertinent data on some of the more sensitive aspects of MJ-12 activities which have been deemed properly classified under the 1954 Atomic Energy [...] of 1954.

I hope this clarifies the necessity to keep current operations with [...]A activities in sensitive areas from becoming official disclosure. From time to time, updates will be provided through NIE as more information becomes available.

(Signed) Allen W. Dulles

This document contains_____ pg

Copy No.   of   copies

Figure 34. Dulles' response to Kennedy's June Memorandum.
Source: Majestic Documents

former decorated US Marine and top-level CIA operative, in his book, *Facing Reality*:

> December 17, Colby informs Angleton that he is relieving him of his two principal duties, his function as Chief of the Counterintelligence Staff and his responsibility for liaison with Israeli intelligence. He gives Angleton the option of remaining in the Agency in a consultant capacity or of retiring before the end of the year…. And Colby gives him two days to reconsider.[275]

On December 25, 1974, Angleton's retirement was announced to the CIA, and the news was quickly leaked to the press. Significantly, his successors soon began a process of burning Angleton's vast file collection. In 1990, Mark Riebling revealed in his book, *Wedge*, that "Angleton's successors had actually burned 99 percent of his CIA files."[276] Apparently, Angleton's files were so sensitive that it was best to simply burn them. It is not surprising that after his death, Angleton's private collection would meet the same fate as those left behind at the CIA after his retirement.

One of Angleton's counterintelligence colleagues, who claimed he was present at the burning of Angleton's files, saved some of the collection. He sent these rescued files to Timothy Cooper, a UFO researcher best known for his role in making public the leaked MJ-12 documents.[277] The partially burned pages of the memorandum were sent to Cooper on June 23, 1999. In the cover letter, the agent states:

> I am a retired CIA counterintelligence officer who worked for Jim Angleton from … [text blacked out] secret files … [text blacked out] sensitive files that would connect MJ-12 to JFK's murder. This document did not exist officially and has never been disclosed within the agency. AWD [Allen

Dulles] was very fearful of disclosure to unauthorized channels and leaks in the White House. I literally snatched the "Directives" from the fire and have kept them safe from review. To allow a review would compromise future directors and put the agency in a difficult position.[278]

According to Dr. Robert Wood and Ryan Wood, the burned document:

... is an original carbon with an Eagle watermark characteristic of government work, but so far forensic laboratories have been unable to trace it.... Although no date is given, its content directly suggests the month of September. The year is estimated to be in the early 1960s and is still under investigation.[279]

The scorched pages date from the Kennedy era and have the characteristics of a government document.[280] If its contents are accurate, it provides smoking gun evidence of the power struggle between Kennedy and MJ-12 over access to UFO information.

The classified Top Secret document with MJ-12 code word access is a set of directives from the Director of the CIA, who simultaneously headed the MJ-12 Special Studies Project, to six other members of the Project. These are identified on the cover page as MJ-2, MJ-3, MJ-4, MJ-5, MJ-6, and MJ-7. It says on the cover page:

As you must know Lancer [Kennedy's Secret Service codename] has made some inquiries regarding our activities which we cannot allow. Please submit your views no later than October. Your action to this matter is critical to the continuance of the group.[281]

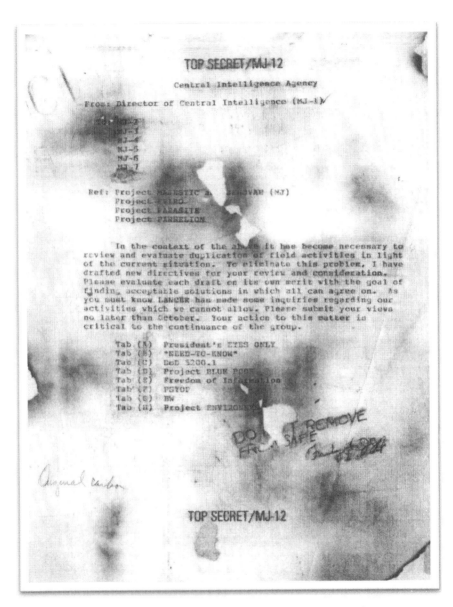

Figure 35. Alleged Top Secret CIA Memo rescued from a fire.
Source: Majestic Documents

The document clearly acknowledges that Kennedy's efforts to gain access to UFO information soon after coming into office on January 20, 1961, actually imperiled the existence of the MJ-12 Special Studies Project/Group.

While the partially burned document pages have no date of issue, the authority of the writer and the political context indicates it was written shortly after Kennedy had issued his June 28, 1961 National Security Action Memorandum requesting a "Review of MJ-12 Intelligence Operations as they related to Cold War Psychological Warfare Plans."[282] The burned document acknowledged that it had "become necessary to review and evaluate duplication of field activities in light of the current situation."[283] This appears to be a reference to the June 28 National Security Action Memorandum review Dulles was ordered to undertake.

The burned document also appears to be a draft for a series of MJ-12 directives from Allen Dulles, who knew his time as CIA Director was limited due to the April 1961 Bay of Pigs fiasco. He needed an answer from the other MJ-12 members by October, a month before he was to retire as Director. Within the burnt pages, the document contained a number of directives concerning how to control UFO information and ensure that it would not be shared with the "Chief Executive [President Kennedy), National Security Council Staff, department heads, the Joint Chiefs, and foreign representatives." Dulles' secret directives proscribed Kennedy's National Security team from gaining access to the most sensitive UFO files possessed by the CIA and MJ-12. These files would have contained much incriminating information about the German space program out of Antarctica, and extensive collaboration at an official level with U.S. authorities.

The most damning directive, drafted by Dulles and apparently approved by six other MJ-12 members, is titled "Project Environment". It is a cryptic assassination directive. In full, it states:

DRAFT

Directive Regarding Project Environment

When conditions become non-conducive for growth in our environment and Washington cannot be influenced any further, the weather is lacking any precipitation … it should be wet.[284]

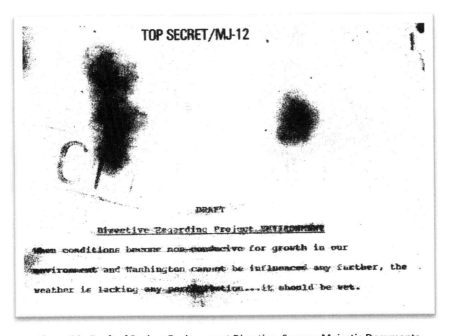

Figure 36. Draft of Project Environment Directive. Source: Majestic Documents

Dr. Robert Wood concluded that this specific page of the burned document is an assassination directive. In an interview discussing it, he pointed out that the cryptic phrase, "it should be wet" originates from Russia where the phrase "wet works" or "wet affairs" denotes someone who had been killed and is drenched with blood.[285]

The code word "wet" was later adopted by the Soviet KGB and other intelligence agencies, according to Dr. Wood. The term "it should be wet", therefore, is a coded command to kill someone. In drafting this cryptic directive, Allen Dulles was seeking approval from six of his MJ-12 colleagues to justify the assassination of any elected or appointed official in Washington, DC whose policies were "non-conducive for growth". The directive is a pre-authorization to assassinate any U.S. President who could not "be influenced any further" to follow MJ-12 policies. The obscure language of the directive insulated the MJ-12 Group in the case of a leak. Its real intent, however, would be clear to any seasoned covert operative. A CIA veteran like James Angleton would know its real meaning and what he was being asked to do once it had been entrusted to him.

While the response of other MJ-12 members to his draft is not found in leaked documents, Dulles' November 5, 1961 letter to Kennedy indicates that his secret draft of MJ-12 directives was approved. Dulles' letter firmly suggests that MJ-12 had decided not to cooperate with Kennedy. Thus, Kennedy's efforts to incorporate MJ-12 psychological warfare activities under the direct control of his National Security Advisor had been dismissed.

Kennedy's failure to pressure the CIA and MJ-12 to yield substantive information about its operations carried an implicit warning. If Kennedy continued in his efforts to demand access to classified UFO/extraterrestrial files and projects, the draft Assassination Directive pointed to the ominous consequences.

Kennedy's Attempt to Cooperate with the USSR on Space and UFOs

In his Inaugural Address on January 20, 1961, President Kennedy indirectly referred to President Eisenhower's farewell speech warning about the growing power of the military-industrial complex. Kennedy described the dangers posed by the armaments industry using science to build ever more destructive weapons:

> Finally, to those nations who would make themselves our adversary, we offer not a pledge but a request: that both sides begin anew the quest for peace, before the dark powers of destruction unleashed by science engulf all humanity in planned or accidental self-destruction.
>
> We dare not tempt them with weakness. For only when our arms are sufficient beyond doubt can we be certain beyond doubt that they will never be employed. But neither can two great and powerful groups of nations take comfort from our present course -- both sides overburdened by the cost of modern weapons, both rightly alarmed by the steady spread of the deadly atom, yet both racing to alter that uncertain balance of terror that stays the hand of mankind's final war....[286]

Kennedy went on to make a bold appeal for cooperation with the Soviet Union in arms control, science and the exploration of space:

> Let both sides, for the first time, formulate serious and precise proposals for the inspection and control of arms, and bring the absolute power to destroy other nations under the absolute control of all nations. Let both sides seek to invoke the wonders of science instead of its terrors.[287]

The most important clue of Kennedy's intention to regain control of the UFO issue and desire to deal with the threat posed by the Fourth Reich is displayed in his appeal for joint cooperation in space with the Soviet Union. During his administration,

Kennedy would repeatedly reach out to the Soviet Union to cooperate in space, along with a host of other areas of mutual concern:

> Together let us explore the stars, conquer the deserts, eradicate disease, tap the ocean depths, and encourage the arts and commerce ... And, if a beachhead of cooperation may push back the jungle of suspicion, let both sides join in creating a new endeavor – not a new balance of power, but a new world of law – where the strong are just, and the weak secure, and the peace preserved. [288]

If cooperation with the Soviet Union were to be established in the international arena, then it would significantly reduce the power of the military-industrial complex. More importantly, cooperation with the Soviet Union would undercut the power of MJ-12 and the Fourth Reich, which had gained exclusive control over the UFO issue, and were using extraterrestrial-related technologies for weapons development.

Kennedy Proposes Joint Space and Lunar Missions with the Soviet Union

In September 1963, President Kennedy launched a groundbreaking initiative to get the USSR and USA to cooperate in joint space and lunar missions. In the background of this publicly announced initiative with powerful Cold War implications was a more secretive attempt by the Kennedy administration to gain access to classified UFO files, and thereby confront the influence of MJ-12 and the Fourth Reich over the U.S. military-industrial complex. Leaked documents reveal that Kennedy instructed the CIA to release classified UFO files to NASA as part of the cooperative space effort with the Soviet Union. If Kennedy had succeeded, there would have been joint space missions to the Moon, and greater sharing of classified UFO files between the CIA, NASA, and the Kennedy administration. This effort would have ensured eventual public release of classified UFO files by both the U.S. and USSR.

In a stunning speech before the United Nations General Assembly on September 20, 1963, President Kennedy said:

> Finally, in a field where the United States and the Soviet Union have a special capacity – in the field of space – there is room for new cooperation, for further joint efforts in the regulation and exploration of space. I include among these possibilities a joint expedition to the moon.[289]

Kennedy was offering to put an end to the space race and start joint missions with the Soviets. According to Nikita Khrushchev's eldest son, Dr. Sergei Khrushchev, this was not the first time that Kennedy had proposed joint space and lunar missions with the USSR. Sergei Khrushchev revealed that at the June 1961 Vienna Summit, less than ten days after Kennedy's famous May 25 speech before a joint session of the U.S. Congress promising to land a man on the moon before the end of the decade,[290] Kennedy secretly proposed joint space and lunar missions to his father. Khrushchev declined, as Sergei Khrushchev later explained: "My father rejected this because he thought that through this the Americans could find out how weak we were, and maybe it would push them to begin a war."[291]

In a series of interviews beginning in 1997, Dr. Sergei Khrushchev said after his father initially refused Kennedy's September 20, 1963 offer of joint space and lunar missions, "in the weeks after the rejection, his father had second thoughts."[292] In one interview, Sergei Khrushchev stated:

Figure 37. President Kennedy addressing United Nations General Assembly
(Sept 25, 1961). Source: JFK Presidential Library

I walked with him, sometime in late October or November, and he told me about all these things. He told me that we have to think about this and maybe accept this idea. I asked why they would know everything, our secrets? He said it's not important. The Americans can design everything they want. It is a very well developed country, but we will have to save money. It's very expensive.... He thought also of the political achievement of all these things, that then they would begin to trust each other much more. After the Cuban missile crisis, his trust with President Kennedy was raised very high. He thought that it's possible to deal with this President, he didn't think that they could be friends, but he really wanted to avoid the war, so through this co-operation they could sojourn their thoughts on these achievements.[293]

Sergei Khrushchev confirmed that his father finally accepted

Kennedy's offer in early November 1963, just over a week before his assassination.[294] According to Richard Hoagland and Mike Bara, authors of *Dark Mission*, the exact date can be traced to November 11, when a key Soviet Mars mission had failed: "A Mars-bound unmanned spacecraft code-named 'Cosmos 21' failed in low Earth orbit exactly one day (November 11) before Kennedy's sudden "Soviet Cooperation Directive to James Webb."[295] Khrushchev's abrupt turnaround, after two years of secret and public overtures by Kennedy, led to a series of immediate Presidential executive actions by Kennedy on the following day.

President Kennedy issued National Security Action Memorandum (NSAM) No. 271 on November 12, 1963. The subject header was "Cooperation with the USSR on Outer Space Matters," and the key passage was:

> I would like you to assume personally the initiative and central responsibility within the Government for the development of a program of substantive cooperation with the Soviet Union in the field of outer space, including the development of specific technical proposals.[296]

The Memorandum furthermore went on to say that the cooperation was a direct outcome of Kennedy's September 20th proposal "for broader cooperation between the United States and the USSR in outer space, including cooperation in lunar landing programs." The Memorandum was classified "Confidential" and addressed to James Webb (NASA Administrator). It was declassified on October 13, 1981.

Significantly, Kennedy added: "I assume that you will work closely with the Department of State and other agencies as appropriate." Kennedy identified the Secretary of State as a key person in implementing the process by which dialogue over the cooperation would take place:

THE WHITE HOUSE
WASHINGTON

November 12, 1963

NATIONAL SECURITY ACTION MEMORANDUM NO. 271

MEMORANDUM FOR

    The Administrator, National Aeronautics and Space
    Administration

SUBJECT: Cooperation with the USSR on Outer Space Matters

I would like you to assume personally the initiative and central
responsibility within the Government for the development of a
program of substantive cooperation with the Soviet Union in the
field of outer space, including the development of specific technical proposals. I assume that you will work closely with the
Department of State and other agencies as appropriate.

These proposals should be developed with a view to their possible discussion with the Soviet Union as a direct outcome of
my September 20 proposal for broader cooperation between
the United States and the USSR in outer space, including cooperation in lunar landing programs. All proposals or sug-
gestions originating within the Government relating to this
general subject will be referred to you for your consideration
and evaluation.

In addition to developing substantive proposals, I expect that
you will assist the Secretary of State in exploring problems of
procedure and timing connected with holding discussions with
the Soviet Union and in proposing for my consideration the
channels which would be most desirable from our point of
view. In this connection the channel of contact developed

UNCLASSIFIED

CONFIDENTIAL

SecDef Control No. X-7448

Figure 38. NSAM 271. Source: Majestic Documents

> I expect you [Webb] will assist the Secretary of State in exploring problems of procedure and timing connected with holding discussions with the Soviet Union and in proposing for my consideration the channels which would be most desirable from our point of view.[297]

This would ensure that the State Department and other U.S. government agencies would have access to the information to be shared with the Soviets under the cooperative space initiative.

In addition to the Confidential National Security Action Memorandum, Kennedy issued a more highly classified "Top Secret" Memorandum to the Director of the CIA, John McCone. Dated the same day of November 12, 1963, the subject header of the file was: "Classification review of all UFO intelligence files affecting National Security". According to a draft of the Top Secret Memorandum that was leaked, Kennedy went on to say:

> [I] have instructed James Webb to develop a program with the Soviet Union in Joint space and lunar explorations. It would be very helpful if you would have the high threat [UFO] cases reviewed with the purpose of identification of bona fides as opposed to classified CIA and USAF sources.... When this data has been sorted out, I would like you to arrange a program of data sharing with NASA where Unknowns [UFOs] are a factor. This will help NASA mission directors in their defensive responsibilities. I would like an interim report on the data review no later than February 1, 1964.[298]

Kennedy's reference to classified CIA and USAF sources of UFO reports show that he was aware that they were systematically separated into classified and unclassified files. The USAF and the other military services were secretly required to

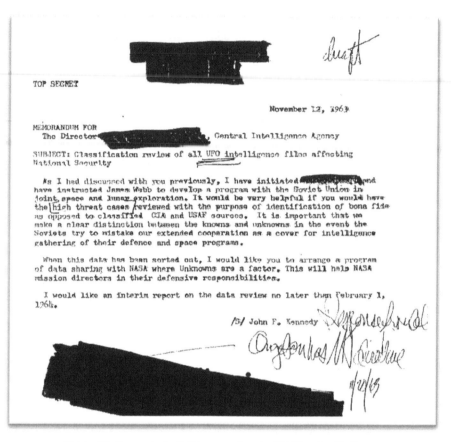

TOP SECRET

November 12, 1963

MEMORANDUM FOR
The Director ████████████ Central Intelligence Agency

SUBJECT: Classification review of all UFO intelligence files affecting
National Security

As I had discussed with you previously, I have initiated ████████████ and
have instructed James Webb to develop a program with the Soviet Union in
joint space and lunar exploration. It would be very helpful if you would have
the high threat cases reviewed with the purpose of identification of bona fide
as opposed to classified CIA and USAF sources. It is important that we
make a clear distinction between the knowns and unknowns in the event the
Soviets try to mistake our extended cooperation as a cover for intelligence
gathering of their defence and space programs.

When this data has been sorted out, I would like you to arrange a program
of data sharing with NASA where Unknowns are a factor. This will help NASA
mission directors in their defensive responsibilities.

I would like an interim report on the data review no later than February 1,
1964.

/S/ John F. Kennedy

Figure 39. Kennedy draft Memorandum to CIA Director McCone.
Source: Majestic Documents

direct their most important UFO files, reported through the CIRVIS (Communications Instructions for Reporting Vital Intelligence Sightings) system created for reporting vital intelligence data by Joint Army Air Naval Publication 146, to the CIA.[299] This is supported by a memorandum by Brigadier General C.H. Bolender on October 1969. He wrote: "reports of unidentified flying objects which could affect national security are made in accordance with JANAP 146 or Air Force Manual 55-11, and are not part of the Blue Book system."[300]

Put simply, there were two sets of UFO files being

collected by the USAF during the Kennedy and later presidential administrations. Those with the least national security significance were made available to the public through Project Blue Book – the "official" public investigation of UFOs by the USAF, which formally ended in 1970.[301] The more important "classified UFO files" which revealed both extraterrestrial activity and the space operations of the Fourth Reich out of Antarctica, were directed into another project that was under the control of the CIA. In particular, the CIA's counterintelligence department controlled access and reported directly to the MJ-12 Group. Requesting the CIA to share UFO files with NASA would in turn lead to its sharing this information with the State Department and other agencies, as stipulated in NSAM 271. Kennedy was, therefore, directly confronting the CIA over its ultimate control of classified UFO files, and attempting to expose the existence of the Fourth Reich's Antarctic space operations.

It is important to note that the Memorandum to the CIA Director refers to the National Security Action Memorandum issued to Webb on the same day. Even though the leaked Top Secret Memorandum to the CIA has not been officially acknowledged (its authenticity has been ranked medium-to-high level) [302], there is no question about the legitimacy of the National Security Action Memorandum (NSAM) 271.[303] NSAM 271 clearly showed that Kennedy had decided to cooperate with the USSR on "Outer Space Matters". If Kennedy had been warned about the dangers of future conflict with the Soviet Union and/or with extraterrestrial life, then sharing classified UFO files was an obvious way to implement NSAM 271.

NSAM 271, and the associated Top Secret Memorandum to the CIA Director issued on November 12, 1963, are evidence that Kennedy firmly linked cooperation with the USSR on "outer space matters" with the release of classified UFO files. Kennedy was aware that the CIA was the lead agency for ensuring the release of classified UFO files, not the US Air Force. Project Blue Book, as many UFO researchers have rightly concluded, was merely a public relations exercise.[304]

President Kennedy's UFO Initiatives Lead to Implementation of Assassination Directive

On November 12, 1963, President Kennedy had reached a broad agreement with Soviet Premier Nikita Khrushchev on joint space missions and sharing classified UFO files. This agreement required both leaders to instruct their respective UFO working groups to share information. Kennedy did this through a November 12th Top Secret memorandum to the Director of the CIA to share UFO files with NASA and the USSR. His memorandum was relayed to James Jesus Angleton, who controlled access to the most highly classified UFO files in the U.S. and was in direct communications with MJ-12, which in turn closely liaised with the Fourth Reich.

Due to Kennedy's request, Angleton followed a Top Secret/MJ-12 set of directives. One of the secret directives, revealed in the leaked and partially burned Top Secret/MJ-12 document (forensically dated to 1961), was the cryptic assassination directive. In case any senior U.S. official did not cooperate with MJ-12, the directive sanctioned political assassination. The leaked document is smoking gun evidence that former CIA Director Allen Dulles was involved in drafting and approving, along with six other MJ-12 members, a treasonous "assassination directive". It is important to emphasize that the MJ-12 "assassination directive" was later implemented by Angleton in direct response to President Kennedy's November 12, 1963 request to the CIA to release classified UFO files.

Kennedy's 1963 efforts to end the Cold War, cooperate with the USSR on joint space missions, and share classified UFO files with the Soviets, created a final showdown with MJ-12. The trigger was Kennedy's fateful agreement with Khrushchev on

November 12, 1963 on space cooperation that led to Kennedy's Top Secret memo instructing the new Director of the CIA, John McCone, to share all UFO information with NASA.[305] Due to NSAM 271, issued the same day, this would ensure that classified UFO files would be shared not only with the USSR, but with the State Department and other U.S. agencies.[306] In short, the two memoranda Kennedy issued on November 12th ensured access to classified UFO files that would extend to more government agencies, ultimately resulting in direct Presidential access. This direct access had been denied to him by McCone's predecessor, Allen Dulles. Dulles had engineered a means by which he could still deny Kennedy access to UFO information, even though he was out of office.

Kennedy's explosive Top Secret November 12, 1963 memo to the CIA Director was relayed by William Colby, then (Deputy) Chief of the CIA's Far East Division, to James Angleton in CIA counterintelligence. It was Angleton who had been given the authority to implement "Project Environment" by the MJ-12 Group if the latter's operations were threatened. The threat that implemented the Directive was the demand by the Kennedy administration for the CIA to release its classified UFO files.

On the bottom of Kennedy's Memorandum to the CIA, next to the signature space, appears the following handwriting: "Response from Colby: Angleton has MJ directive 11/20/63." Colby is acknowledging that Angleton, two days before Kennedy's assassination, had the MJ directive – the burned document – and would use it to respond to Kennedy's Memorandum. This handwriting directly implicates the MJ-12 Group and Angleton in the Kennedy assassination due to the cryptic MJ-12 assassination directive.

Therefore, the assassination of President Kennedy was the direct result of his efforts to gain access to the CIA's control of classified UFO files. Unknown to Kennedy, a set of secret MJ-12 directives issued by his former CIA Director, Allen Dulles, ruled out

/S/ John F. Kennedy

Figure 40. Handwriting at bottom of Nov 12, 1963 Kennedy Memorandum to CIA Director McCone: "Response from Colby – Angleton has MJ directive. 11/20/63"
Source: Majestic Documents

any cooperation with Kennedy and his National Security staff on the UFO issue. It was Dulles and another six MJ-12 Group members who sanctioned the directives found in the burned document, including a vague political assassination directive against non-cooperative officials in the Kennedy administration. This could be applied to Kennedy himself if the official entrusted to carry out the MJ-12 Assassination Directive concluded that the President threatened MJ-12 operations.

While Dulles and his six associates pre-authorized the assassination of any political figure who threatened MJ-12 Group operations in late 1961, it would not be implemented until later in the Kennedy Administration. The Assassination Directive had been passed on to Dulles' close ally James Jesus Angleton, the CIA counterintelligence chief, for safekeeping and possible execution. Even though he would no longer be CIA Director, Dulles had engineered a means whereby he would still be able to deny Kennedy access to the CIA's classified UFO files – and even deprive Kennedy of his life – if he demanded access to them.

It was Kennedy's joint space cooperation initiative with the

USSR and the demand that the CIA share all UFO information with NASA, the State Department and the Soviets that triggered the execution of the assassination plan. Kennedy's November 12, 1963 Memorandum to CIA Director McCone for the CIA to share UFO information was judged to be a direct threat to MJ-12 Group operations and risked exposing the Fourth Reich's secret space program out of Antarctica. Colby's handwritten reference to Angleton having the directive is very significant. It reveals how Angleton, in his official capacity as head of the CIA's counterintelligence division and safe-keeper of the classified UFO files, was authorized to respond to any UFO ultimatum by the Kennedy administration.

Angleton consequently made the decision to go ahead with the implementation of the Assassination Directive according to the classified instructions he received when the Directive was entrusted to him by Dulles in late 1961. The Directive then was approved by the MJ-Group to whom Angleton ultimately answered. The Assassination Directive had been cryptically written, thereby insulating the MJ-12 Group from possible blowback in the case of a leak. A seasoned covert operative like Angleton knew its real meaning. It was Angleton who gave the orders for assembling a CIA hit team to assassinate President Kennedy in accord with a set of cryptic instructions he had received in late 1961 by the MJ-12 Group, which liaised closely with the Fourth Reich in setting policies designed to maintain secrecy over extraterrestrial life and secret space programs.

Direct support for the CIA involvement in the Kennedy assassination due to his efforts to share classified UFO files with the Soviet Union comes from former CIA operative E. Howard Hunt. Hunt is best known for his role in the Watergate burglary of the Democratic National Headquarters at the Watergate Hotel. Hunt's trial and conviction captivated Washington, DC, and became known as the infamous Watergate Scandal.

In one of the Nixon tapes, the disgraced former President discussed Hunt's importance insofar as he had information that could blow open what really happened in the Kennedy Assassination. Nixon told his Chief of Staff, H.R. Haldeman:

> [V]ery bad, to have this fellow Hunt, ah, you know, ah, it's, he, he knows too damn much and he was involved, we happen to know that. And that it gets out that the whole, this is all involved in the Cuban thing, that it's a fiasco, and it's going to make the FBI, ah CIA look bad, it's going to make Hunt look bad, and it's likely to blow the whole, uh, Bay of Pigs thing which we think would be very unfortunate for CIA and for the country at this time, and for American foreign policy, and he just better tough it and lay it on them.[307]

In his memoirs, Haldeman later revealed that the "Bay of Pigs" was used by Nixon as a code word for the Kennedy assassination: "It seems that in all those references to the Bay of Pigs, he was actually referring to the Kennedy assassination."[308] Haldeman's admission clearly suggests that in Nixon's view, Hunt was directly involved in the Kennedy assassination.

In his famous "last confession" to his son, Saint John, in 2007, Hunt confirmed his involvement as a "bench warmer" for a CIA hit team planning the assassination.[309] Saint John Hunt discussed the taped confession in the April 5, 2007 edition of *Rolling Stone Magazine*.[310] What is even more momentous, however, is what Howard Hunt told his Watergate legal advisor and friend, Douglas Caddy, about the reason for the JFK assassination.

In a November 2017 interview with veteran UFO researcher, Linda Moulton Howe, Caddy for the first time publicly revealed what Hunt had confidentially confided to him.

> And I said, "Howard, you told me about going into these Cuban government reports at the Democratic

104-10120-10356

13042

BIOGRAPHIC PROFILE (PART 1) SCD: 7 Sep 1944

NAME (Last, First, Middle)			SEX	DATE OF BIRTH	LONGEVITY COMP. DATE
HUNT, E(verette) Howard			M	Oct 1918	8 Nov 1949

MARITAL STATUS	DEPENDENTS (Incld. sm. Slaves)	NO. (Times) OF BIRTH				US NATURALIZATION DATE(S)	
Married		5	1920 1951 1952 1954 1963			NA	NA

CAREER STAFF STATUS	MEMBERSHIP	OTHER STATUS		LAST MO. RPT. QUAL. FOR		EVAL. FOR
	Jul 1954			Feb 1967	TDY Standby	TDY Standby

RESERVE STATUS	NONE SERVICE	GRADE	ACTIVE DUTY WITH CIA CAT.-1	RELEASE TO MIL. SER. CAT.-1	IN RE DEFERRED DEFERRED CAT.-3
X					

ASSESSMENT DATE	PROFESSIONAL TEST DATE	LANGUAGE APTITUDE TEST DATE
None	None	None

NON-CIA EMPLOYMENT
40-42   Military Service, US Navy, Ensign
42-43   "The March of Time," NYC - Script Writer
43   "Time," Inc, NYC - War Correspondent (South Pacific, 9 mos)
43-46   Military Service, USAAF, 1st Lt (1945-46, OSS in China)
46-49   Free Lance Writer
48-49   Economic Cooperation Administration, Paris, France - US Media Specialist

NON-CIA EDUCATION 1944 AAFSAT, Orlando, Fla - Air Combat Intelligence (4 mos)
36-40   Brown Univ - AB, English, English Literature, Economics
50   Berlitz School of Languages, DC - Spanish

FOREIGN LANGUAGE ABILITIES (Waived, Preliminary, Date Trained)
Spanish - R,P Inter; W,S,U High (Apr 1967) Transl & Interpr -May 1957
German - R,W,S,U Slight;P,intor; T,none - May 1957 (declined testing)
French - R,P Elem; W,S,U Slight; T None - Sep 1966 - disc prof Apr 1968

AGENCY SPONSORED TRAINING
50 Admin Proc   1953 Photography
50 Secret Writing
53 Ops Famil
53 Flaps & Seals

CIA EMPLOYMENT HISTORY SINCE 18 SEPT 1947 (Personnel Actions, Military Orders, and Criminal Details)

EFFECTIVE DATE	POSITION TITLE & OCCUPATIONAL CODE	GRADE	AC	ORGANIZATION & ORGAN. TITLE (If any)	LOCATION
v 1949	I.O. (Editor) 0130.00	13		OPC/P&P Stf/Program Grp II	Hq
c 1950	I.O. 0132.00	13		OPC/Latin America/Ops/COS	
n 1951	" 0132.00	14		OPC/Latin America/DCOM	"
g 1953	Ops Off 0132.00	15	PP	DDP/SE/Ch, PP Staff	Hq
n 1954	Ops Off (PP) 0136.31	15	DP	DDP/FE/SR-NA/Ch, PP Staff	"
b 1957	Area Ops Off 0136.01	15	DP	DDP/WH-II/ COS	
v 1960	Ops Off 0136.01	15	D	DDP/WH-4	Hq
v 1961	Jun-Nov 1961 Detailed to Office of			DCI	
1962	Ops Off 0136.01	15	D	DDP/(Astf/Plans&ResGrp/Ch, EvalBr	"
1962	" " 0136.01	15	D	DDP/CA Staff/OC	"
1962	" " 0136.01	15	D	DDP/DODS/Facilities Br/Ch, W.PSec	"
ug 1964	" " 0136.01	15	D	DDP/DOD/U.S. Field/Ch, CA Staff	"
eb 1965	" " 0136.01	15	D	DDP/Off of the DDP/Ops Group	"
	Jul 1965-Sep 1966 Contract Employee				
ep 1966	Ops Off 0136.01	15	D	DDP/WH/Operations Stf	Hq
on 1967	" " 0136.01	15	D	DDP/Eur/Spec Act Stf	"
ug 1968	" " 0136.01	15	D	DDP/Eur/Ops Staff	"
n 1970	Retired				

DATE REVIEWED	BIO PROFILE REVIEWED BY	ITEMS 1-18 REVIEWED & VERIFIED BY EMPLOYEE

Figure 41. E. Howard Hunt Military & CIA Service Record

National Committee and the Kennedy assassination. What was in the reports? Why was Kennedy killed? Why were those reports so important?"

Howard Hunt said, "Kennedy was killed because he was about to give our most vital secret to the Soviet Union."

And I repeated, "Our most vital secret? What would that have been?"

Howard leaned forward and looked in my eyes and he said, "The alien presence." And he shook my hand and walked away.[311]

Caddy's disclosure of what Hunt had confided is extraordinary since it provides powerful corroboration that Kennedy's plan to cooperate with the Soviets in gaining access to the CIA's classified UFO files was the trigger for his assassination.

In assessing the assassination of President Kennedy in light of his attempt to gain access to classified UFO files that would expose the existence of the Fourth Reich's Antarctica based space program, it would be fair to conclude that his death was a combination of well-known domestic U.S. factors and a previously unknown international factor. Joseph Kennedy had made deals with both the Italian Mafia and the Fourth Reich to get his son elected. However, President Kennedy along with his brother, Attorney General Robert Kennedy, pursued policies that antagonized both the Mafia and the Fourth Reich. Kennedy threatened to expose not only the German Antarctica space program, but its infiltration of the U.S. military-industrial complex.

The MJ-12 Group which had the responsibility for maintaining the secrecy system, laid the foundation for Kennedy's removal through the eight directives it gave to the CIA's counter-intelligence chief, James Angleton, in 1961. Angleton was therefore the intermediary for the confluence of powerful domestic and foreign elements that felt Kennedy had betrayed them, was a threat to future operations, and thus wanted him dead. The Kennedy Assassination marked both the death knell of the American Republic and the sinister influence of the Fourth Reich over all aspects of U.S. life, which extends to our current era.

CHAPTER 9

Siemens Implicated in Tracking Forced Labor & Slaves in Space

Siemens' History of Slave Labor

Europe's largest engineering company today, Siemens AG, was the major corporate contractor for the construction of key components and prototypes for spacecraft developed secretly during World War II in Nazi Germany and Antarctica. Throughout that turbulent period, Siemens already loomed as the logical choice for secretly supporting the industrial infrastructure for the German Antarctica space program due to its ingenuity, leadership and industrial prowess.

During WWII, and immediately after, Siemens was headed by Herman von Siemens, a Ph.D. in physical chemistry who was the Chairman of Siemens & Halske AG (founded 1847), and its sister company, Siemens-Schuckertwerke AG (founded 1903). These two, and a third Siemens' offshoot, Siemens-Reiniger-Werke (founded 1932), were combined in 1966 to form the present day Siemens AG. It currently employs over 350,000 people worldwide, and in fiscal year 2016 generated over US$80 billion in revenue.[312]

After Nazi Germany's defeat, Herman von Siemens was

detained on December 5, 1945 for questioning over the Siemens companies use of slave labor. According to the public record:

> He was brought to the Nuremberg trials as a prisoner to deal with war crime charges, but finally no prosecution was filed, as there were no personal misdeeds traceable. The charges were dropped, so he could return as head of company in 1948.[313]

It is certain that von Siemens would have been heavily interrogated over his knowledge of the Siemens companies' participation in the building of spacecraft for the Nazi and Antarctica programs. As noted earlier, it is all but certain that Siemens companies had established subsidiaries in Antarctica, and moved a lot of equipment and resources down there. It is very plausible that his cooperation was a factor in his release without trial to resume leadership of the Siemens companies.

Despite von Siemens release without charges being filed, it is undisputed that the Siemens companies he headed extensively used slave labor in their many factories hidden all over Nazi Germany. It took decades, but on September 24, 1998, Siemens AG decided to begin compensating victims of its former slave labor practices as the following Associated Press Report described:

> Siemens announced plans Wednesday for a $12-million fund to compensate former slave laborers forced to work for the firm by the Nazis during World War II.... Almost a year ago, at its 150th anniversary celebrations, the company had insisted that it could do no more for its former slave laborers than express "deepest regrets." The Munich-based Siemens said its fund is in addition to the $4.3 million it paid to the Jewish Claims

Conference in 1961 and to providing humanitarian help for victims. Siemens estimates that between 10,000 and 20,000 slave laborers worked in its wartime factories.[314]

Siemens' role in acknowledging its wartime use of slave labor and efforts to compensate victims was a step in the right direction, but the size of the compensation fund, when compared to the up to 20,000 people abused, is shamefully astounding. Their acknowledgement, coming forty years after the events in question, leads to troubling questions over whether the company was sincere or simply wishing to avoid class action lawsuits then underway in the United States, which promised to award far more to the former slave laborers:

> The threat of lawsuits has raised the pressure on German firms to pay direct claims to the thousands of concentration camp inmates, mostly Jews, forced to work in their factories. Lawyers representing former slave laborers criticized Siemens, as they did Volkswagen, for setting up a fund to avoid larger payments a lawsuit might demand. Siemens is seeking "the cheapest alternative," said Munich attorney Michael Witti, who with a colleague filed the U.S. lawsuit. [315]

Readers might be forgiven for believing that the compensation of former slave laborers is an issue long past and only associated with a terrible chapter in our recent history. However, there is credible whistleblower testimony that such practices continue and Siemens is still involved.

Siemens Covertly Builds Billions of RFID Chips that can Track Slave Labor

William Pawelec was a computer operations and programming expert with the US Air Force, who started his own electronics security company, and worked for high profile U.S. defense contractors such as SAIC and EG&G. He received high-level security clearances and had access to many classified projects. Prior to 2001, Pawelec decided to reveal what he knew about deep black projects that were hiding advanced technologies from the U.S. public. He gave an interview to Dr. Steven Greer with the strict instruction that it would only be published after his death. He died on May 22, 2007 and the video was published posthumously on December 14, 2010.[316]

Among his many revelations was information concerning the development of the first electronic RFID tracking chips that were developed as early as 1979. Pawelec explained the history of their development, and the role his Denver-based company played in setting up meetings with government agencies, which were interested in using the chips for security purposes. In his video interview with Dr. Greer, Pawelec said:

> At the time in the security industry, a lot of us had a lot of concerns about tracking and locating people that had been kidnapped. Particularly what was going on in Europe at the time where we were having NATO officers, even the Prime Minister of Italy, kidnapped ... These people were drained [of information] or they were brutalized or both ... One of the goals of the industry was to develop technology that would allow us to track these people or locate them quickly. Hopefully to save their lives but on a secondary basis to keep from being drained of sensitive information.[317]

He further explained that the tracking chips, which were very small and shaped like a pill, had multiple functions:

> Now this particular pill shaped device, very minute, had a lot of flexibility in its capabilities. It was basically almost a transponder. You could send a frequency to it and it would respond back with its unique number which could not be changed once the chip was made. Yet there were a lot of capabilities that could be added to this chip such as monitoring temperature, blood pressure, pulse, and even wave forms out of the brain.[318]

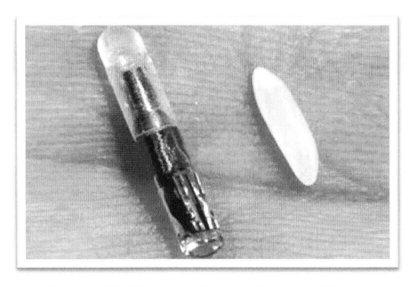

Figure 42. RFID Chip compared to grain of rice. Source: Wikipedia

Soon after demonstrating an even more sophisticated lithium niobate chip in 1984, which could be tracked from a distance of 120 kilometers in space, Pawelec said he discovered a small Silicon Valley company had been set up to manufacture billions of them. He learned that "after they had made billions and billions of these little chips" the factory was shut down a year later, and all information about the chips disappeared.[319]

The small company responsible for making the billions of tiny chips, according to Pawelec:

> [W]as a division of a rather major European electronics firm that had the plant. Siemens. What concerned me was that they had built these chips and who knows what happened to them, and they built them in the billions in volume, because they are so small that you could take a six-inch wafer and make hundreds of thousands of them on a wafer, and they've disappeared somewhere.[320]

Pawelec's statement here is very significant since it shows how a giant conglomerate like Siemens can easily set up U.S. subsidiaries to build products useful to the German secret space program in Antarctica, and then dissolve the subsidiary in a way that is very difficult to track and investigate.

Pawelec went on to describe how his colleague, "Bob", the former head of security for the U.S. State Department, was assassinated in Nairobi, Kenya, because he was getting too close to learning about the people who had gained control over the tracking chip technology and had built billions of them for an unknown purpose. According to Pawelec, the people behind Bob's assassination had infiltrated the U.S. military-industrial complex at its highest level, and could intimidate and silence anyone:

> Bob was killed and it was a hit, and it's always concern me today that he had gotten a little too close to who had been involved with this implantable chip technology we've been trying to, for couple years then, quietly trying to find out who had been doing it without our government realizing it was going on. Whoever it is has got total

ability to penetrate anytime anywhere our government and locate what is going on instantly. Research since the early eighties on my own and with some friends indicates that we have at least four power groups in the world. They have wealth beyond all imagination, they have advanced technologies, they have taken over various programs, particularly black programs, within our government and probably even the Russian government, and the Chinese. Politics to them, as we know it, is not the same; and they have agendas totally unlike what ... we perceive our government's agenda is really are and that they are able to track unbelievably what's going on around at a minute level.[321]

Here, Pawelec alludes to the Fourth Reich, which had begun infiltrating the upper echelons of military and corporate power in the U.S. since the Eisenhower administration.

Pawelec's testimony suggests that the Siemens corporation had acquired the rights and control over the tracking chip technology, built billions of them in less than a year using a U.S. subsidiary, and then arranged for the local manufacturing plant to close with all information vanishing about the tracking chips. Importantly, the Siemens corporation was associated with a powerful force embedded within the U.S. military-industrial complex, which was intent on hiding the true purpose of the RFID chips. This hidden force possessed the power to remove anyone who got too close to learning the truth, even the head of security at a U.S. Embassy.

It is worth repeating that the main function of the tiny chips, according to Pawelec, was to track people and even monitor their key physiological processes over large distances. This leads to the question; why would the Siemens corporation need billions of chips to monitor people over large distances? The answer that emerges from what has already been discussed is that the RFID tracking chips were needed for the slave laborers used in Antarctica and in the space colonies established by the German breakaway civilization.

The Antarctic German colony had an extensive number of bases that used slave labor, and likely exported captured humans as slaves for off-planet use as well. The unique individual identifiers within the RFID chips would provide a means, for whoever had purchased or acquired the slave labor, to be able to monitor their human assets using advanced satellite surveillance systems. Support for such a disturbing conclusion comes from William Tompkins and Corey Goode, whose 'insider testimonies' have revealed many details of the German secret space program operating out of Antarctica.

In a nutshell, both Goode and Tompkins claim that the Antarctica based Germans infiltrated the U.S. military-industrial complex using Operation Paperclip scientists. As discussed earlier in this book, thousands of German scientists and engineers were fast tracked into senior positions in the U.S. military-industrial complex. Major U.S. corporations had been infiltrated along with many leading aerospace and engineering companies in Europe, Japan and elsewhere.[322]

The Siemens company, with its long history of being a major armaments supplier for the German military, was a natural conduit for further penetration of the U.S. military-industrial complex. Pawelec had learned about the strange influence exercised over the U.S. military-industrial complex by German elites when he traveled to the Tonopah Test Range for a classified project. After Pawelec's death, his wife, Mary Joyce Annie DeRiso, shared more of the information that Pawelec had told her about the Tonopah incident, and who was really in control of the U.S. military.

DeRiso shared her information in an interview she gave about Pawelec's testimony which was presented in Greer's Disclosure Project video:

His disenchantment began when he was called to give what he thought was going to be a regular project status report at Tonapah. The meeting was held in a heavily controlled room that was built like a Faraday cage making it impossible for communications to come in or out of the sealed room. Briefcases, papers, pagers and any form of identification were not allowed at that meeting ... Only the generals could be recognized by their uniforms. The tension was really high and Bill was surprised at how nervous the high-ranking generals were. He knew something BIG was up. Bill saw a private jet escorted by two of our military jets land on the tarmac. Surprisingly, this private jet rolled all the way to the building where the meeting was scheduled as the escort jets departed. A very imposing man stepped out of the jet and entered the room. He was relatively tall, and wore a very expensive European suit. His shoes and briefcase were equally luxurious and there was an aide or bodyguard by his side. His demeanor was very aristocratic and he spoke with a High German accent. The room was electrified with nervous tension as each person gave his status report and answered questions.

When everyone had spoken, the German man thanked them for their good work and simply left. He was never introduced nor identified in anyway. It is believed he was Baron Jesco von Puttkamer, one of the Germans who came to the United States with Werner von Braun. Whatever happened that day convinced Bill that the United States, and probably the whole world, was being controlled by Europeans ... but exactly who 'they' were was the big question.

It drove Bill and his friends on a quest to find out what was really going on. After that, he frequently quoted his friend Jim Marrs who often says, 'The Nazis may have lost the battles but they won the war.'[323]

Based on deRiso's interview, Pawelec believed that remnants of the Nazi regime had survived WWII, and that German elites were now in control of Western Europe and the United States.

The testimony of Pawelec's widow is markedly significant since it reveals her husband's final conclusion over who really controls the U.S. military-industrial complex. Notably, this conclusion matches what Corey Goode and Bill Tompkins later revealed. Many talented German scientists had been brought into the U.S. under Operation Paperclip, and fast tracked into senior leadership positions in NASA, U.S. corporations, and the military industrial complex more generally. This finally takes us to the question of what purpose the Siemens company had for building billions of chips for tracking humans from 120 kilometers in space?

Siemens and the Galactic Slave Trade

In a *Cosmic Disclosure* interview, William Tompkins explained how the Nazis had used slave labor during WWII, transferred them to Antarctica and today still use slaves:

Germany had massive underground facilities that were all [using] slaves and even to the extent that when the decision was made before the war ended that they were going to continue all of their extraterrestrial developments on UFOs and on every weapon system that they were building, they took the production facilities to Antarctica, but they also brought the slaves with them. So now there are slaves underneath the ground and they still are today in Antarctica.[324]

Tompkins went on to describe a galactic slave trade:

But the slave business out there is a big business, and this is happening today. It's not something that happened 100 years ago. This has been going on a long time and that needs to be fixed. There's many different classes of people that are abducted for slavery, sexual slavery. They want the top and the smartest, because they are worth more.

They have, I think, four or five different levels of people that they abduct. They abduct top medical research people. They abduct the corporate levels, and they abduct the most brilliant levels, and then they go down through the three levels and that says where they get sectioned off.

Everybody gets to two planets and then it's decided where they're going to be sent to. But it's a massive business. It's been going on for years, and we can't identify where these people have gone. Just like 'normal' abductions, we don't know where they went, because most of them don't come back. We're only hearing from the few that came back.[325]

In an email interview, Goode outlined the development of a galactic slave trade that involved extraterrestrials and national elites secretly in control of Earth governments and militaries. He said this was achieved through a network of corporations, called the Interplanetary Corporate Conglomerate (ICC):

The Secret Earth Governments and their Syndicates discovered that a large amount of humans were being taken off the planet by various ET's anyway, so they decided to find a way to profit from it and have control over which people were being taken. In prior arrangements they were made promises of receiving technologies and biological specimens for allowing groups to abduct humans, but the ET's rarely delivered on their promises. Once they had developed the advanced infrastructure (ICC) in our Sol System, along with advanced technologies (that some of the thousands of ET groups traveling through our system were now interested in obtaining) and now had the ability to deter most unwelcome guests from entering Earths airspace, the Cabal/ICC then decided to use human trafficking as one of their resources in interstellar bartering.[326]

Tompkins and Goode's revelations provide an answer to *why* the Siemens company chose to build billions of trackable chips with unique identifiers. It is highly likely these chips were used to monitor the galactic slave trade, as well as forced labor in Antarctica, and in secret colonies on Mars and elsewhere within our solar system. Satellites or spacecraft could track RFID embedded humans, thereby providing the German space program intelligence on where and how slave labor was being used.

It has already been established that in 1998, Siemens AG acknowledged its involvement in the Nazi practice of using slave labor and agreed to compensate those forced to work in its companies. Siemens publically announced the estimated number to be between 10,000 to 20,000 victims.[327] If the testimonies of Pawelec, Tompkins and Goode are accurate, then the number of victims of a continuing practice of secret forced labor and slave trade in space have become incredibly larger. By tracking of victims in an illicit galactic slave trade, through tiny RFID chips, the Siemens corporation is complicit. Eventually, Siemens and any other corporations involved in exploiting forced labor in secret space colonies or a galactic slave trade will eventually have to confess to their involvement, face justice, and compensate victims of these vast undisclosed crimes against humanity. This will only occur after the truth emerges about events in Antarctica, and when the secrecy system implemented there is overcome.

CHAPTER 10

The Antarctic Treaty & Keeping the German Space Program Secret

Key Elements of the Antarctic Treaty

On June 23, 1961, the Antarctic Treaty went into force after being ratified by its 12 original signatory nations: Argentina, Australia, Belgium, Chile, France, Japan, New Zealand, Norway, South Africa, the Soviet Union, the United Kingdom and the United States. Currently, there are 50 nations party to the Treaty, including those with major aerospace programs such as Brazil, China, Germany, India and Italy.

The Treaty stipulates that Antarctica is to be used for peaceful purposes to advance human knowledge. The stationing of military forces and all forms of weapons development and testing are outlawed under article 1 of the Treaty:

> 1. Antarctica shall be used for peaceful purposes only. There shall be prohibited, inter alia, any measures of a military nature, such as the establishment of military bases and fortifications, the carrying out of military maneuvers, as well as the testing of any type of weapons.[328]

In addition to the permanent demilitarization of Antarctica, articles 2 and 3 outline the extent to which Treaty signatories will cooperate in scientific exploration:

> Article II
> Freedom of scientific investigation in Antarctica and cooperation toward that end, as applied during the International Geophysical Year, shall continue, subject to the provisions of the present treaty.
>
> Article III
>
> 1. In order to promote international cooperation in scientific investigation in Antarctica, as provided for in Article II of the present treaty, the Contracting Parties agree that, to the greatest extent feasible and practicable:
>
>> (a) information regarding plans for scientific programs in Antarctica shall be exchanged to permit maximum economy and efficiency of operations;
>>
>> (b) scientific personnel shall be exchanged in Antarctica between expeditions and stations;
>>
>> (c) scientific observations and results from Antarctica shall be exchanged and made freely available.[329]

Verification is critical for any successful international treaty, since nations have to be sure that their strategic rivals are not taking advantage of cooperative nations behind closed doors. Article VII describes how contracting nations can send observers to any base in Antarctica to ensure full compliance of the Treaty provisions:

Article VII

1. In order to promote the objectives and ensure the observance of the provisions of the present treaty, each Contracting Party whose representatives are entitled to participate in the meetings referred to in Article IX of the treaty shall have the right to designate observers to carry out any inspection provided for by the present Article. Observers shall be nationals of the Contracting Parties which designate them. The names of observers shall be communicated to every other Contracting Party having the right to designate observers, and like notice shall be given of the termination of their appointment.[330]

Article IX outlines how future meetings of Treaty signatories could act in ways to further the Treaty's key goals:

(a) use of Antarctica for peaceful purposes only;

(b) facilitation of scientific research in Antarctica;

(c) facilitation of international scientific cooperation in Antarctica;

(d) facilitation of the exercise of the rights of inspection provided for in Article VII of the treaty;

(e) questions relating to the exercise of jurisdiction in Antarctica;

(f) preservation and conservation of living resources in Antarctica. [331]

The Secretariat for the Antarctic Treaty was established in Buenos Aires, Argentina. The *U.S. Antarctic Guide* explains the functions of the Treaty's Secretariat:

The Antarctic Treaty consultative parties established a secretariat in Buenos Aires, Argentina, for support of Antarctic Treaty activities. Besides assisting with preparation for annual meetings, the Secretariat also is responsible for information related to the Treaty System and the Protocol.[332]

The Treaty was heralded as a major step forward for turning Antarctica into a vast nature preserve to be studied in a peaceful way for the benefit of all humanity. Here is how the *U.S. Antarctic Guide* explained this alleged achievement:

> The treaty is a remarkable achievement whose primary success has been to reserve the area south of 60 degrees South latitude as a zone of peace: it prohibits measures of a military nature, including fortifications, and it prohibits nuclear explosions and the disposal of radioactive waste. It gives treaty parties the right to inspect all areas of Antarctica, including stations, installations, equipment, ships and airplanes of other member states, to ensure continuing adherence to the treaty.[333]

Similarly, a 1969 *New York Times* editorial stated:

> [T]he Antarctic Treaty helped to create foundations of mutual confidence on which great diplomatic landmarks were based, notably the test ban treaty of 1963, the space compact of 1967, and the nuclear nonproliferation pact of 1968. Later thinkers saw Antarctic Treaty influence on the 1979 Moon treaty and the 1982 Law of the Sea convention.[334]

However, at the core of the Antarctica Treaty a major lie was hidden, known only to a few of its major signatories. While the signatory nations pledged to maintain Antarctica as a demilitarized zone dedicated to peaceful development and scientific exploration, already present deep inside the continent's interior resided a flourishing German space program that stood in violation of all the key elements of the Antarctic Treaty. Major military installations were built in Antarctica by the German space program during World War II, and advanced weapons research has continued at a vigorous pace since the war's end, deep below the ice shelf unknown to the world public.

What had fatally undermined the Treaty was that the Fourth Reich, which had successfully established safe havens in Antarctica and South America after WWII, was neither acknowledged nor recognized by the international diplomatic system as a state. This meant that German bases in Antarctica were under no clear international legal authority since Germany had been divided in two after the Nazi defeat. On February 5, 1979, its western half, the Federal Republic of Germany, gave their consent to be bound by the Antarctic Treaty.[335] However, neither German state was in any position to assert authority over the Fourth Reich bases in Antarctica. As far as the Antarctic Treaty signatories were concerned, the hidden German bases were ignored, and an international pretense that they simply didn't exist was maintained.

Furthermore, Buenos Aires, where the Secretariat for the Antarctic Treaty was established, was a well-known center for Nazi exiles, including Martin Bormann.[336] As previously presented, the US Army Intelligence file titled the "Red House Report" exposed how Bormann had started preparations for a post-War Nazi economic revival from locations such as Argentina. Adolf Hitler had joined with Bormann in Argentina as a number of official government documents and eyewitness testimonies have indicated.[337] Argentina, through the high-level Nazis who escaped there, was the unofficial capital of the Fourth Reich. This meant that the Fourth Reich, through its covert presence in Argentina, would effectively be able to monitor and manipulate the Antarctic Treaty Secretariat to achieve its goals.

The way in which the Antarctic Treaty undermined U.S. dominance and military operations in Antarctica was noted by some U.S. Senators who strongly denounced ratification, as explained in an article in the *Antarctica Sun*:

> "I rise in opposition to the ratification of this treaty" echoed in the chambers. The U.S.S.R. had signed it and couldn't be trusted, the United States had failed to take possession of territory despite "solid claims to some 80 percent of the Antarctic," we would forfeit future economic potential, and nuclear explosions were banned. "We are trading what I would call a horse for a rabbit," one Senator said, "to get the concessions the treaty would grant in the way of international amity and accord."[338]

Compounding the situation further were the U.S. corporations that began working as military contractors for the German Antarctica space program as a result of the Eisenhower administration agreement. Operation Paperclip scientists, who had thoroughly penetrated the U.S. military-industrial complex and NASA, could identify any new promising technologies or products and then have them secretly built for the German space program. These would either be secretly shipped to Antarctica, or assembled there. What Siemens AG did through a U.S. shelf company in building billions of RFID chips was a pattern that would repeat itself around the world.

Essentially, while signatory nations would build bases on the surface of Antarctica and commit themselves to the Antarctic Treaty articles, the German run space program would not be restrained by any of the Treaty's provisions. As a non-signatory entity in Antarctica, which was ignored by Treaty signatories, the

Fourth Reich was free to conduct advanced military-related research and development without troublesome observer inspections. This made the Fourth Reich's underground Antarctica facilities an attractive partner for major U.S. and European armaments companies interested in advanced weapons research and development.

U.S. and European corporate contractors that conducted operations at the German Antarctic facilities were able to pursue advanced military research in ways that were unfettered by Antarctic Treaty provisions. Any kind of military research that gave a qualitative edge to the German run space program could be expected to flourish, with major support by U.S. and European corporations. The slave labor policy that was adopted for the German Antarctica facilities, for example, provided abundant human subjects for advanced genetic experiments and bio-weapons research. In an upcoming chapter, I will discuss alarming claims that hundreds of thousands of slaves have been sacrificed in large scale biological experiments.

Formation of the Interplanetary Corporate Conglomerate

After the 1955 agreement was reached with the Eisenhower administration, there slowly emerged an international consortium of corporations building all the key components for craft to be used in the German Antarctica space program. By the 1980's, this international consortium had its own fleets of spacecraft making up a powerful corporate run space program, which operated alongside the German-run program based deep beneath the Antarctica ice shelf.

Corey Goode has described the original German Antarctica space program as the "Dark Fleet", since its operations were largely unknown to the space program being run by the US Navy – Solar Warden.[339] The corporate run program that emerged was called the Interplanetary Corporate Conglomerate (ICC), and it quickly rose as a powerful rival to the Navy's Solar Warden program, which became operational in the early 1980's according to both William Tompkins and Goode.[340]

The ICC was a fusion of the key German companies that had first established operations in Antarctica during WWII, along with the U.S., European and other corporations that began collaborating with the Antarctic based Germans. Over time, this led to large industrial bases being built in Antarctica where the ICC could conduct its weapons research, build advanced craft for the Dark Fleet (which primarily operated outside of our solar system), and its own separate space fleet that was used by the ICC for its operations on Mars and elsewhere in the solar system. About the ICC, Goode wrote:

> The ICC has an entire industrial infrastructure that includes bases, stations, outposts, mining operations and facilities on Mars, various moons and spread throughout the main asteroid belt (where a "Super Earth Planet" once existed). They have facilities to take raw materials and turn them into usable materials to produce both complex metals and composite materials that our material sciences have not dreamt of yet. They have separate groups of facilities that produce various types of technologies as well as each facility or plant that produces a specific component of a technology so that those working in the facilities and living in the support colonies/bases do not know exactly what they are producing. Much of the time the components are multiuse and are used in cross over projects. There are facilities on Earth [e.g. Antarctica] that operate in much the same manner that contribute to the SSP on several levels.[341]

The key to ensuring activities deep below the Antarctic ice shelf would remain publicly hidden was to keep the world media and general public out of Antarctica, and strictly monitor all scientific research by Antarctic Treaty signatories so the scientific community, conducting legitimate environmental research, did not learn the truth. A final step was to keep any discoveries of ancient artifacts secret to guarantee that no one learned about Antarctica's extensive ancient cavern system and the artifacts retrieved from ancient civilizations.

These became high priority policies for the Dark Fleet and the ICC. In this way, their Antarctic operations remained hidden for decades within a frozen continent that would persist as an enigma for the world public. The secrecy policy had dramatic effects for any nations or intrepid explorers who wanted to open Antarctica up to the rest of the world. One of the Antarctica Treaty's original signatory nations, New Zealand, found out how high the cost to be paid was by countries wanting to open up the continent for international tourism.

Air New Zealand Flight TE 901 Antarctica Crash

Beginning in February 1977, Air New Zealand began offering tourist flights from Auckland, with a stopover in Christchurch, to Antarctica's Ross Ice Shelf. The route took tourists over various island groups, a flyby of the massive volcano, Mt. Erebus, and a loop around McMurdo base. The flights proved quite popular and by 1979, four flights were offered. Then on November 28, 1979, disaster struck.

Air New Zealand flight TE 901 crashed into Antarctica's Mt. Erebus, killing all 257 passengers and crew. The New Zealand Airline Pilots Association gave details of how Flight TE901 mysteriously flew off course and directly into the mountain:

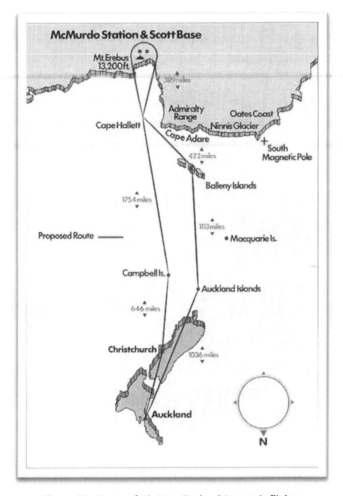

Figure 43. Route of Air New Zealand Antarctic flights

On the night before the flight, flight planners made what they thought was a small correction to an earlier mistake made some months previously, when the flight plans were computerized. In their mind, they were only shifting the final Antarctic waypoint about two miles, which was about the expected error usually found in flights of a similar duration. Now, according to the airline witnesses, the navigation staff always knew the track ran more or less directly over Mt. Erebus, and the shifting of the waypoint some two miles would still run the track more or less over the volcano. But in fact, what they had done was to shift the route from McMurdo Sound, to over Mt. Erebus, a change of nearly 30 nautical miles.[342]

The initial investigation concluded that pilot error was responsible, but a public outcry over the findings led to a Royal Commission being convened. It was conducted by the highly respected justice, Peter Mahon, QC. A summary of Mahon's findings included:

... the coordinates in the navigation computer had been changed without telling the crew or the flight followers at Mac Center. At the time of the crash, TE901 was flying in local whiteout conditions (in clear air under cloud cover, but with no surface definition), but most of the flight had been in clear air.... As it was, the flight crew was confident of their position and flight path until the collision alarms sounded just before the crash.[343]

Mahon concluded that Air New Zealand executives had conducted "an orchestrated litany of lies" to evade the airline's responsibility for the disastrous course correction made to the flight, and instead steered blame towards the pilot for the crash.[344] Mahon's findings proved very controversial and led to the Prime Minister of New Zealand, Robert Muldoon, challenging him to name the "conspirators and liars".[345] Mahon's report led to a successful appeal by Air New Zealand, and his damaging findings were overturned. Controversy continues to this day over ultimate responsibility for the crash, and Mahon's findings.[346]

Mahon's conclusion that there had been a conspiracy, "an

orchestrated litany of lies" by the airline to hide the truth behind the crash, strongly alludes to the real culprits responsible for the tragedy. The course correction, made without informing the flight crew, was not merely incompetence, but in reality a case of sabotage by forces not wanting tourist activities occurring in Antarctica. The ICC, with its vast global corporate resources, including embedded assets in major airlines such as Air New Zealand, is the most likely candidate for the mysterious chain of events that led to the TE 901 crash into Mt. Erebus.

Up to 1994, when Australia's Qantas Airlines began tourist flights once again over Antarctica, tourist flights did not occur, thereby minimizing damaging sightings of Antarctica's many anomalies.[347] Throughout the entire recent history of Antarctica, however, there have been regular military flights occurring, and sometimes a brave crew member takes the risk and comes forward to reveal some of the sighted anomalies.

US Navy Flight Engineer Reveals Anomalous Events in Antarctica

A retired 20 year US Navy flight engineer publicly came forward in 2015 to reveal some of his experiences in Antarctica. He was stationed there from 1983 to 1997, and flew over 4000 hours as flight engineer for the Antarctic Development Squadron Six (aka, VXE-6). He wrote a letter to veteran UFO researcher Linda Moulton Howe and used only his first name "Brian" to tell his story, since he currently works with a major corporate contractor. He supplied various documents to Howe including his military discharge papers (DD214), and consented to several public interviews with her where he appeared as a highly credible source giving details about ongoing events in Antarctica.[348] In 2016, I briefly met Brian, and later, together with freelance journalist, Kathryn Leishman, tracked down a former Antarctic worker who importantly recalled Brian from a 1984 flight squadron yearbook, and thereby confirmed that he was genuine as a whistleblower.

Figure 44. A Lockheed LC-130F Hercules ski-equipped VXE-6 at the Amundsen–Scott South Pole Station. Image courtesy of Brian.

In his letter to Howe, Brian described three anomalous events, each of which casts further light on several topics already raised in earlier chapters. He supplied a map of Antarctica with red crosses marking the locations of each of the incidents he described in his letter (see Figure 46).

In one incident that occurred in the 1985/86 polar season, "Brian" was ordered to transport a sick scientist from Australia's Davis Base to McMurdo base in Antarctica. In flying from McMurdo, Brian and his crew flew directly over a restricted air space of the South Polar base, Amundson Scott. While doing this, he witnessed a large hole going directly into the ice and extending deep into the interior:

> Another unique issue with South Pole station is that our aircraft was not allowed to fly over a certain area designated 5 miles from the [Amundsen-Scott] station. The reason stated because of an air sampling camp in that area.
>
> This did not make any sense to any of us on the crew because on 2 different occasions we had to fly over this area. One time due to a medical evacuation of the Australian camp called Davis Camp.
>
> It was on the opposite side of the continent and we had to refuel at South Pole and a direct course to this Davis Camp was right over the air sampling station.
>
> The only thing we saw going over this camp was a very large hole going into the ice. You could fly one of our LC130 into this thing.
>
> It was after this medevac mission where we [were] briefed by some spooks (Intelligence Agents I presumed) from Washington, DC and told not to speak of the area we overflew.[349]

Brian described that the hole looked as though it was naturally formed, and not artificially made with equipment.[350] This is very significant since recently scientists have confirmed the effect of heightened volcanic activity under the ice caps, which is causing the surface ice to move up and down by melting the ice under the surface.[351] This raises the possibility that volcanic activity is causing the holes on the surface as escaping heat rises all the way up through miles of ice. More about this phenomenon will be explored later in a chapter presenting the testimony of Corey Goode concerning secret excavations of an ancient pre-flood civilization under the Antarctic ice sheets.

In a November 2017 radio interview, Brian added that the opening of the South Pole hole appeared to be descending like a ramp into the interior, rather than a steep vertical drop.[352] He and other crew members saw snow mobile marks going into the hole from the Amundson-Scott station about five miles away. This implies that an operation was underway, whereby equipment and personnel were being moved back and forth from the station into a location somewhere inside the hole.

Brian also responded to a question by Howe about the chatter he was picking up about a joint human-extraterrestrial (EBE) base located in the area near the South Pole and the large hole:

> We are told not to talk among ourselves officially. But the guys after a flight, you have a few beers and it's like, 'I heard these scientists talking about that there's some guys there at Pole that were working with these strange-looking 'men.' They weren't saying, you know, 'alien' or 'extraterrestrial,' or whatever. And that the air sampling station was actually a joint base with the scientists and the E.T.'s.[353]

Brian emphasized that he, himself, never directly heard what scientists were saying about the joint base, and that it comes second-hand from his flight crew.[354] While these are only second-hand reports, they do help corroborate that agreements had been reached in the 1950's, whereby the U.S. government started to work with a German-Reptilian alliance in Antarctica.

In the next incident, Brian described repeatedly seeing during the 1995/96 season, silver disc-shaped craft flying over the

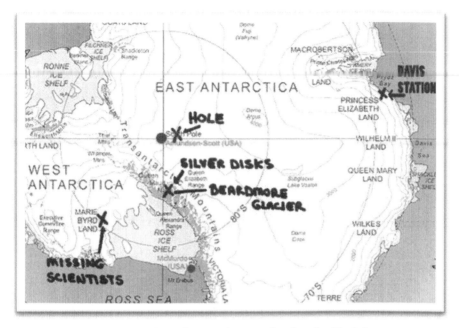

Figure 45. Locations of Antarctic anomalies described by Brian.

Transantarctic mountain range:

> Between these two stations [McMurdo and Davis] is a mountain range called the Trans Antarctic's. With what we called Severe Clear weather from McMurdo to South Pole the Trans Antarctic's are visible from the altitudes which the aircraft flew approximately 25,000 to 35,000 feet.
>
> On several flights to and from South Pole our crew viewed air vehicles darting around the tops of Trans Antarctic's almost exactly in the same spot every time we would fly by and view them.
>
> This is very unusual for air traffic down there due to the fact that the only aircraft flying on the continent were our squadron aircraft. Every aircraft

knew where the other aircraft were due to flight schedules being followed.[355]

The significance of this incident is that it reveals that Antarctica is host to at least one secret base possessing saucer-shaped craft similar to the Vril and Haunebu vehicles developed by Nazi Germany, which were reported in the 1947 attack against Admiral Byrd's Operation Highjump.

The final incident involved a group of around 15 scientists that went missing while doing 3 weeks of field work in Marie Byrd Land, Antarctica during the 1994/95 season as Brian best recalls. The scientists were out of radio communications for a two week period when Brian's team was sent out to investigate and found them missing from their base, as he explained in his letter:

> One outlying camp (near Marie Byrd Land) we dropped scientists and their equipment it was out of communication with McMurdo for 2 weeks. Our crew flew back to the camp to find out if the scientists were ok. We found no one there and no sign of any foul play.
>
> The Radio was working fine as we called McMurdo to verify it working properly. We left the camp and flew back to McMurdo as ordered by our CO. A week later the Scientist showed back up to their camp and called McMurdo for someone to come pick them up.
>
> Our crew got the flight back there to pick them since we put them into that camp and we knew the terrain and location. None of the scientists would talk to any of the crew on the plane and to me they looked scared.

As soon as we landed back at McMurdo they (Scientists) where put on another of our squadron aircraft and flown to Christchurch New Zealand. We never heard about them again.

Their equipment that we brought back from the camp was put in quarantine and shipped back to the United States escorted by the same spooks that debriefed us about our flying over of the air sample camp/ large hole in the ice.[356]

Brian's statement that the scientists were not at their camp when he was first sent to investigate, suggests that they had discovered or were taken into the interior of Antarctica. It's worth noting that their location was in the vicinity of one of the buried volcanoes that is linked to large caverns formed out of thermal activities. Brian emphasized that the scientists were scared and wouldn't talk.

In his November 2017 interview, Brian said the scientists appeared to be suffering from PTSD, and that after he and his flight crew had returned the scientists' equipment back to Christchurch, New Zealand, they were again debriefed to remain silent about what they saw at McMurdo.[357] Brian says that he later heard from another aircrew that after his aircrew's return of the scientists' equipment to McMurdo, it was sent to a base in Ohio. He agreed with Howe that the equipment was most likely taken to Wright Patterson Air Force Base for evaluation.

Conspicuously, in 2016, after Brian met with researcher Linda Moulton Howe to discuss the missing scientists and other anomalies, he received a threatening phone call. He detailed this conversation to Howe:

"And the voice on the other end of the line said, 'Is this Brian?'

I said, 'Yes, who is this?' because I didn't recognize the number. 'I want to tell you that what you have been talking about you need to stop talking about.'

And I said, 'Well, what stuff was that?'

And he says, 'We know that you were with Linda Howe on last Thursday night and we know that you went to dinner at a Mexican restaurant in Joshua Tree and we were aware of what you were talking about with her and other people and that your experience that you had when you were on the ice when you were in the service, we don't want you talking about that. Specifically, we don't want you talking about the scientists that you picked up after being missing for a couple of weeks.'

And I said, 'How would you know about that? I've only talked to certain people.'

And the voice in the phone said, 'Well, we know pretty much everything.' And he said, 'Just don't be putting that out there anymore! Certain people would prefer that you not talk about that.'

And I said, 'Well, I'll consider that.'

And then the phone went dead. There was a click. Whoever was on the other end hung up. It was like, 'Wow, this is out of the blue!'"[358]

When he investigated the phone number of the caller, he discovered it was the general number of the National Security Agency out of Fort Meade, Maryland.

The missing Antarctic scientists had apparently discovered something important, but it scared them in some way. They were forbidden from sharing what they saw with the plane's crew, and whatever they brought back with them, including their equipment, that was taken to the U.S. What did the scientists see that alarmed them, and why was Brian warned off by the NSA from discussing them? And what precisely is the human-extraterrestrial connection in Antarctica?

CHAPTER 11

Lake Vostok Mystery & Ancient Ruins in Antarctica

Lake Vostok Magnetic Anomaly

In 1957, the Soviet Union built a base in eastern Antarctica in a region they named "Vostok" (Russian for East). Arguably, the Russians chose their base site very carefully, since it later turned out to be at the tip of a remarkable subterranean lake. The existence of the lake had been hypothesized as early as 1959 by Russian geographer Andrey Kaptisa, but was not scientifically confirmed until 1993 using orbital laser altimetry.[359] The Russians named the vast body of water Lake Vostok, after their base.

Here is how the *Antarctic Sun*, which is published out of McMurdo base for the United States Antarctic Program, described events leading up to Lake Vostok's discovery:

> When the Russians opened Vostok Station near the geomagnetic pole in 1957, they had no idea that it was situated over an ancient body of water more than 1,640 feet (500 m) deep and 243 miles (230 km) long. And when they started drilling the world's deepest ice core in an attempt to

understand recent global warming in relation to
the climactic cycles of the last 500,000 years, they
would not have predicted that they would be
stopped at 11,886 feet (3,623 m) by a group of
scientists concerned with contamination of the
lake's pure water.... Although early seismic surveys
in the 1960's and 70's indicated that water might
exist under the ice cap, it wasn't until drilling was
well under way in the early 1990s that satellite,
seismic, and airborne radar data were put together
to map the buried lake. "It was a 'Eureka!'
moment," said Martin Seigert, a University of
Bristol glaciologist.[360]

Lake Vostok measures 250 km (160 mi) by 50 km (30 mi) at its
widest point, and by volume is one of the largest lakes in the
world.

Various scientific missions have been conducted to unravel
some of the mysteries behind Lake Vostok. In 1998, the Russians
drilled to a depth of just over 100 meters above the lake and took
ice core samples, which showed the existence of extremophile
microbes leading scientists to conclude that the lake contained
life. This in turn led to NASA considering Lake Vostok an ideal
place to develop sterile drilling and robotic probe technologies
that could be used on future missions to Jupiter's moon, Europa.

The *Antarctic Sun* wrote about the ensuing plans to those
already underway by NASA, which did not appear to present any
insuperable technological challenge:

The next phase could involve NASA tests of the
robots. The cryobot would melt its way down to
the lake where it would eject the hydrobot to
explore the depths and send back pictures and data
to the surface via a cable. The final stage would
involve deep coring to retrieve sediment and water

samples. The details of probing the lake without introducing contaminants are still being worked out. It is a complex and ambitious effort that with the help of NASA technology will potentially answer some fundamental questions about the evolution of life here on Earth. And by giving scientists a testing ground for the cryobot and the hydrobot, something may someday be discovered about the evolution of life on other planets.[361]

NASA's plans were also covered by Britain's *Telegraph* newspaper on September 21, 1999:

> Lake Vostok is likely to be the oldest of all the "subglacial" ice lakes because of its size. If it has been isolated for 40 million years, there would have been enough time for unique creatures to evolve, as opposed to creatures that have adapted to a new environment.
>
> The Antarctic studies may be a prelude to similar missions elsewhere in our solar system, notably to Jupiter's moon Europa. NASA regards the Vostok mission as a test-bed for the search for alien life on the oceans thought to exist on Europa.
>
> The Vostok exploration would take place in the next five years. The exploration of Europa would be in a series of missions beginning in 2003 and lasting for 15 years.... The first entry of a probe into Lake Vostok will require extraordinary precautions to ensure that the vehicle and its instruments are clean, so as not to contaminate the pristine lake.
>
> One suggestion is to use a Cryobot, a 10ft 6in pencil-shaped device with a heated tip that

unspools a cable carrying power and a fiber-optic video and data cable.

The Cryobot splits into two under the ice and the top half stays at the ice-water interface to hunt for life. The lower part (the point of the pencil) continues down a smaller cable until it hits the sediment at the bottom, where it will also search for life and release a Hydrobot, a tiny submarine equipped with sonar and a camera. The Hydrobot rises like a soap bubble, reporting what it sees above and below it.[362]

Grants were awarded to the Jet Propulsion Laboratory by the National Science Foundation to send a probe into Lake Vostok by 2002, using its new sterile drilling technologies in order not to contaminate the pristine environment.[363]

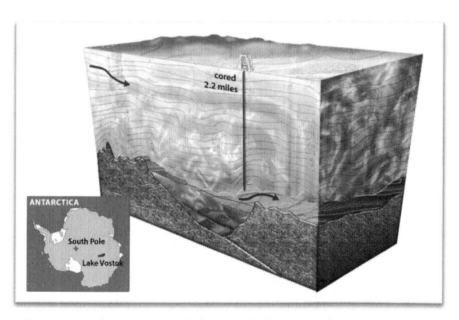

Figure 46. Artist's cross-section of Lake Vostok drilling. National Science Foundation.

In the Antarctic summer of 2000/2001, the Support Office for Aerogeophysical Research sent 36 flights over Lake Vostok and used ground-penetrating equipment to map the Lake. Here is how Kristan Hutchinson Sabbatini, writing in the February 4, 2001 edition of the *Antarctic Sun*, described the mapping process:

> Every second the equipment recorded the gravitational attraction, six radar readings and 10 measures on the magnetometer. The altimeter gave the altitude of the ice to within 10 to 20 centimeters. Radar showed the terrain below the flat ice changed from rolling plains on one side of the lake to mountains on the other. The lake itself appeared to be in a basin, below two miles (three to four km.) of ice.[364]

Detailed maps were created for the first time showing the subglacial Lake Vostok by a Columbia University team led by Dr Micheal Studinger.[365] The 2001 mapping survey displayed something else that attracted scientific attention and sparked great public interest – a large magnetic anomaly. Sabbatini wrote:

> The evidence is a huge magnetic anomaly on the east coast of the lake's shoreline. As the first SOAR flight crossed over to the lake's east side, the magnetometer dial swung suddenly. The readings changed almost 1,000 nanotesla from the normal 60,000 nanoteslas around Vostok. A tesla is the standard measure of magnetism. Studinger typically finds anomalies of 500-to-600 nanotesla in places where volcanic material has poured out of the ground. "When we first saw this huge magneticanomaly, that was very exciting," Studinger said.

Usually magnetic anomalies are much smaller and it takes some effort to distinguish the anomaly from normal daily changes in the magnetic field. In this case there was no confusion.

"This anomaly is so big that it can't be caused by a daily change in the magnetic field," Studinger said.

The anomaly was big in another way, encompassing the entire Southeast corner of the lake, about (65 by 46 miles) 105 km by 75 km. The size and extremity of the magnetic anomaly indicated the geological structure changes beneath the lake, and Studinger guessed it might be a region where the earth's crust is thinner. [366]

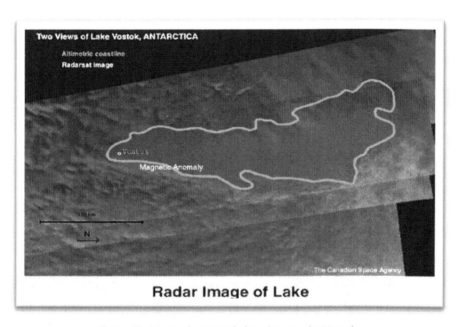

Radar Image of Lake

Figure 47. Magnetic Anomaly location at Lake Vostok

After the February 2001 discovery of the giant magnetic anomaly, plans very quickly changed for NASA, the National

Science Foundation (NSF) and other organizations over sending robotic probes down into Lake Vostok. By the end of 2001, the NSF was talking about a decade-long delay out of environmental concerns, as the November 18 edition of *The Antarctic Sun* reported:

> The NSF has scrapped a timeline sketched out at an NSF workshop in 1998, which would have had access holes drilled into the lake next year and samples removed in 2003.
>
> That schedule has been delayed as much as 10 years, said Julie Palais, glaciology program manager for the Office of Polar Programs.
>
> "Anyone who thinks about this realistically will realize it's going to take a long time to develop the technology," Palais said. "To me it's one of the most challenging projects I've ever been involved in as far as the how-tos."[367]

Was it merely environmental concerns that led to the cancellation of plans to send robotic probes into Lake Vostok, or was there another reason for the abrupt change in plans?

Was an Ancient City Discovered at Lake Vostok?

Two veteran researchers of NASA programs, Richard Hoagland and Mike Bara, said there was far more to the magnetic anomaly than just a long term geological process related to the Earth's crust being thinner in this section of Antarctica, thereby leading to the magnetic anomaly. They wrote in a May 2001 article about an alternative explanation to the one originally portrayed by Dr. Studinger:

Others, like *Enterprise* consulting geologist Ron Nicks, have serious difficulty with this theory. Nicks explains that such a thinning would heat the underlying rock and thus *diminish* (rather than increase – as observed) the crust's ability to locally amplify the Earth's magnetic field.

There is, as always, an equally viable alternative explanation. An anomaly like this could also be caused by an accumulation of *metals* – the kind you would get if you found *the ruins of an ancient, buried city!*

An "ancient city under the ice?" Such a discovery would be absolutely dazzling, sending shockwaves through our world as profound as the discovery of "artifacts on Mars" or "ruins on the Moon." And the notion is not as improbable as you may think.[368]

The discovery of an ancient city or some other large artificial object buried under two miles of ice would certainly have been earth shattering news. Such a discovery would explain why NASA/JPL and the National Science Foundation suddenly abandoned the idea of sending robotic probes into Lake Vostok with cameras, which would have publicly televised what lay hidden there.

Hoagland and Bara further described how JPL's stated reason for pulling back on its robotic probe plans suddenly became an internet controversy involving the National Security Agency:

Almost immediately after the discovery of the Columbia "Vostok magnetic anomaly," word began to leak out that JPL was inexplicably "pulling back from its Vostok exploration program." The reason given was the previously stated "environmental concerns." This was all well and good, until unconfirmed reports began to surface that a JPL spokesperson had admitted at a February press conference that the National Security Agency (NSA) had literally taken over the JPL polar research program at Lake Vostok. It was this report which created something of a firestorm on the Internet.[369]

The NSA's surprising involvement in the Lake Vostok anomaly gained the attention of other researchers such as Henry Stevens, who wrote:

> The reason for the NSA's involvement had nothing to do with biology and everything to do with physics…. [T]he fact remains that the best and most probable answer as to the origin of this vast magnetic disturbance is the presence of a massive amount of metal. Metal as in a buried city…. Is this metal the remains of Atlantis? Is it the underground city said by some to have been built by the Germans, Neu-Berlin? Is this an extraterrestrial base? With the NSA involved, the only thing for certain is that we won't be told the answer to this mystery anytime soon… [370]

Hoagland and Bara described a series of unprecedented medical evacuations from Antarctica in mid-2001, and the important fact that some of them were Raytheon Corporation employees, indicating that a "black project" had been initiated to explore the Lake Vostok anomaly:

> [S]ome "Special Project" has, against all scientific and environmental prudence, indeed drilled

through the ice into the Lake Vostok eco-system (clandestinely, of course). And, the participants have suddenly found themselves exposed to "something" for which their bodies literally have no immunity — something not extant in the rest of Earth's biosphere for between 13,000 and several million years! After the initial reports of "four emergency extractions," the number changed to five ... and now *twelve* McMurdo personnel are supposedly in need of a dangerous, "emergency medical evacuation" well into the Antarctic winter season. At one level, this has all the earmarks of "something" virulently spreading among the limited winter population at the Base, something that even the fairly complete medical facilities at McMurdo can no longer cope with. Complicating the picture is the fact that the "extractees" are not research scientists or long-term support personnel, but are all employees of Raytheon Corporation — a high-tech firm that is deeply involved in a variety of black-ops programs for the U.S. government all around the world.[371]

Alleged Missing @tlantis TV Crew

On April 13, 2002, a press release appeared on the *Atlantis Mapping Project* website claiming that a videographer crew from "@lantis TV" had gone missing after filming a massive archeological discovery. The press release boldly emphasized the "Spectacular Ruins" captured on video, and the efforts by @lantis TV to recover the confiscated video footage.

U.S. Denies Spectacular Ruins in Antarctica Captured on Video
MP LA 03-18-02 0925GMT ^ | Monday, March 18, 2002 - Web posted at 5:25 a.m. EDT (0925 GMT) | WASHINGTON, D.C. (AMP)

osted on 4/13/2002, 4:01:52 PM by vannrox

U.S. Denies "Spectacular Ruins"
n Antarctica Captured on Video

WASHINGTON, D.C. (AMP) -- The U.S. government said it will seek to block the airing of a video found by Navy rescuers in antarctica that purportedly reveals that a massive archeological dig is underway two miles beneath the ice. The @lantisTV roduction crew that shot the video is still missing.

ttorneys for Beverly Hills-based @lantisTV stressed that the company's primary concern is for the safety and welfare of its crew. ut they stated they will "vigorously oppose" any attempts to "censor material that is clearly in the public interest and public omain." The ice continent of Antarctica, they point out, belongs to no nation. The U.S. has no jurisdiction there.

Figure 48. News Release about alleged Antarctic Video

However, an investigation of the "Antarctic Mapping Project" and @lantisTV revealed that both were merely a marketing gimmick for a then upcoming book by Thomas Greanias titled *Raising Atlantis,* which was first published in July 2005. Indeed at the bottom of the Press releases by the "Antarctic Mapping Project", the following disclaimer appeared.

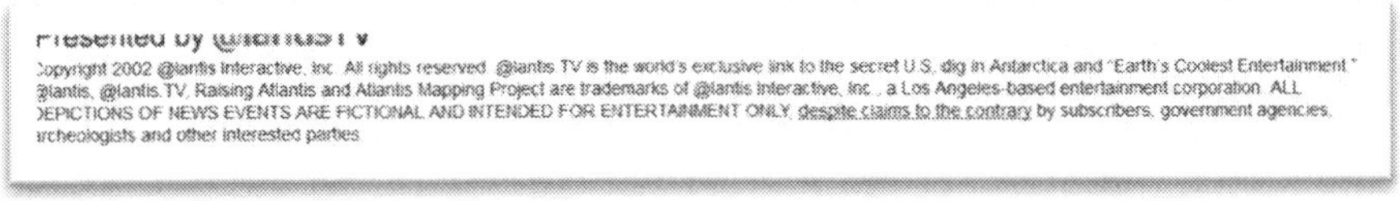

Presented by @lantisTV

Copyright 2002 @lantis Interactive, Inc. All rights reserved. @lantis TV is the world's exclusive link to the secret U.S. dig in Antarctica and "Earth's Coolest Entertainment." @lantis, @lantis.TV, Raising Atlantis and Atlantis Mapping Project are trademarks of @lantis Interactive, Inc., a Los Angeles-based entertainment corporation. ALL DEPICTIONS OF NEWS EVENTS ARE FICTIONAL AND INTENDED FOR ENTERTAINMENT ONLY, despite claims to the contrary by subscribers, government agencies, archeologists and other interested parties.

Figure 49. Disclaimer Statement

The disclaimer clearly states that all of their press releases were purely fictional and for entertainment only, thus there was no film footage of spectacular ancient ruins found in Antarctica or a missing television crew. Was the mysterious press release merely a clever marketing gimmick, or was their some truth to the mysterious excavations and missing crew?

To this day, various major media outlets and blogsites mistakenly refer to the 2002 press release, and subsequent *Antarctic Mapping Project* blogsite references to it, as genuine.

For example, on December 12, 2016, Jennifer Hale from Britain's *Sun* newspaper wrote:

> Conspiracy theorists went wild earlier this year when a video claiming to be from the lost city emerged. It appeared to show extensive ancient ruins hidden in the ice, and was a video supposedly 'left behind' by a California TV crew who have been missing since 2002.
>
> Archaeologist Jonathan Gray claimed that the US government is trying to block the video from being seen because it reveals there is a "massive archaeological dig under way two miles beneath the ice".[372]

Why has a fictitious press release about an alleged 2002 ancient Antarctic ruins discovery proved so resilient over the years? An explanation worth consideration is that it is a "psychological operation" to hide the truth in plain sight. Typically, such intelligence psy-ops reveal the truth, but in such a way that is easily discredited.

Let's revisit the earlier incident recounted by the Navy flight engineer Brian, who told about the missing Antarctic scientists, and was warned off by the NSA about discussing them further with Linda Moulton Howe. Was the NSA and the U.S. intelligence community trying to misdirect the general public from genuine discoveries made of ancient ruins by planting false stories in the public sphere about missing videographers who allegedly filmed an ancient city?

Curiously, on February 3, 2012, Russian scientists drilling deep into the Lake Vostok region of Antarctica reportedly went missing according to a *Fox TV* news story:

The world holds its breath, hoping for the best after six days of radio silence from Antarctica – where a team of Russian scientists is racing the clock and the oncoming winter to dig to an alien lake far beneath the ice.

The team from Russia's Arctic and Antarctic Research Institute (AARI) have been drilling for weeks in an effort to reach isolated Lake Vostok, a vast, dark body of water hidden 13,000 ft. below the surface of the icy continent. Lake Vostok hasn't been exposed to air in more than 20 million years.

The team's last contact with colleagues in the unfrozen world was six long days ago, and scientists from around the globe are unsure of the fate of the mission – and the scientists themselves – as Antarctica's killing winter draws near.[373]

The missing scientists' story was quickly dismissed as a mere misunderstanding, and it was announced five days later that the Russians had finally succeeded in reaching Lake Vostok on February 8, 2012.[374] Was an ancient lake the only thing discovered by the Russians? The falsified rumors of missing scientists may well have been part of yet another psychological operation to create confusion about what was really discovered at Lake Vostok and/or elsewhere under the Antarctic ice sheets.

Discovery & Excavations of Ancient Ruins in Antarctica

Corey Goode states that he first heard about an advanced civilization in Antarctica that had been flash frozen from a senior officer, who he dubbed "Sigmund". This officer was part of a USAF/DIA/NSA/NRO secret space program investigation into

Goode's claims surrounding a highly advanced program, Solar Warden, run by the Navy. Sigmund led a covert mission that involved multiple abductions and debriefings of Goode who was being tested for the fidelity of his information.[375] After Sigmund was satisfied about the accuracy of Goode's information and sources, he unexpectedly shared some of his knowledge about the Antarctica excavations. Goode says more information was forthcoming during a military abduction that took place on October 26, 2016, when Sigmund revealed information about his own activities, including time spent in Antarctica:

> Furthermore he told me that he had been stationed at several military installations in Antarctica and had spent time in the very area where the Anshar had taken me on a reconnaissance flight.[376]

Sigmund went on to describe recent discoveries in Antarctica, which explained the growing scientific, political and religious interest over the frozen continent:

> He [Sigmund] stated that an extremely ancient series of cities had been discovered flash frozen deep under the ice-shelf. He confirmed that there were also many animals and "Pre-Adamites" preserved in the ice…. They were all flattened/ crushed or knocked over by the event that flash froze the area. They have tons of trees/ plants and wildlife frozen in place, like they were put on pause.[377]

If true, this would not only confirm rumors about a major discovery at Lake Vostok, but would also be startling confirmation of the research on the Earth's crustal displacement conducted by Sir Charles Hapgood. His studies produced evidence of pole shifts in the past that have led to the Earth's axis of rotation shifting dramatically over a short period, where even sub-tropical areas might suddenly find themselves at the poles. Hapgood's 1958 book, *Earth's Shifting Crust*, featured a foreword by Albert Einstein endorsing the rigor of Hapgood's research.[378] Einstein summarized his colleague's theory as follows:

> Polar wandering is based on the idea that the outer shell of the earth shifts about from time to time, moving some continents toward and other continents away from the poles. Continental drift is based on the idea that the continents move individually ... A few writers have suggested that perhaps continental drift causes polar wandering. This book advances the notion that polar wandering is primary and causes the displacement of continents.... This book will present evidence that the last shift of the earth's crust (the lithosphere) took place in recent time, at the close of the last ice age, and that it was the cause of the improvement in climate." [379]

Hapgood's thesis that the last pole shift happened at the end of the last ice age, approximately 11,000 BC, could now be remarkably verified by the discovery of a flash frozen Antarctica civilization. Furthermore, this discovery would also confirm that the Oronteus Fineus map, which shows an ice-free Antarctica, is based on ancient historical records which divulge Antarctica in fact once possessed a thriving civilization before a devastating Pole Shift event.[380]

Goode has offered some information about activities surrounding the Antarctica discovery:

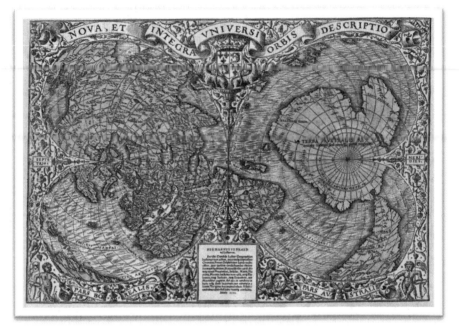

Figure 50. Oronteus Fineus map shows Antarctica ice free

The first discoveries occurred sometime back, not sure. They had an idea of what was below the ice after doing very high-tech scans from space. They had been excavating one site and discovered many, many others spread out across what used to be dry ground. This happened over time.

They have been studying what they are finding, and moving some of it out before bringing in various bigwigs from various secret societies. Then they do tours like the "Cabal Disneyland" they have in the stasis chamber in Ohio. They are continuing to excavate, but know what is around in the areas where they have cleared away the ice.

They are worried that all of the steam excavation could cause the surface of the ice shelf to collapse down on the sites they have uncovered. I think they would have to go public in some sort of a way to go bigger on this excavation and widen it. [381]

Goode has also provided an artistic sketch of the ancient archeological discovery (see figure 51), which shows the ruins being accessed through a vast hole into the ice, where snowmobiles and tractor-trailers are able to descend down one side. This is significant since it corresponds to what Brian reported seeing during his overflight of the South Pole hole in 1985/86, where vehicle tracks were visible going from the nearby Scott-Amundson base into the hole via a ramp down into the interior.

A discovery of this magnitude is clearly Earth-shattering news for the archeological community, and helps to explain why prominent world figures such as U.S. Secretary of State John Kerry, Astronaut Buzz Aldrin and Sir Peter Cosgrove, Australia's Governor General, traveled to Antarctica in 2016 – to see the discoveries first hand.[382] The visit by Russian Orthodox Church Patriarch Kirill is perhaps the most significant because it suggests that a discovery was made near a major Russian Antarctic station, Lake Vostok. Regarding this stream of recent VIP visitors to Antarctica, Goode asserts: "The finds in Antarctica are a major reason the World Political/Religious leaders have been brought down there to tour what has been found in the last year." [383]

Goode also offers a description of the inhabitants of this ancient Antarctic civilization, as relayed to him by the high-ranking officer, Sigmund:

> He described the "pre-Adamites" as beings with elongated skulls, with strangely proportioned bodies that were obviously not designed for Earth's gravity and atmosphere pressure. This group had apparently arrived here from another planet in our solar system that was no longer hospitable.

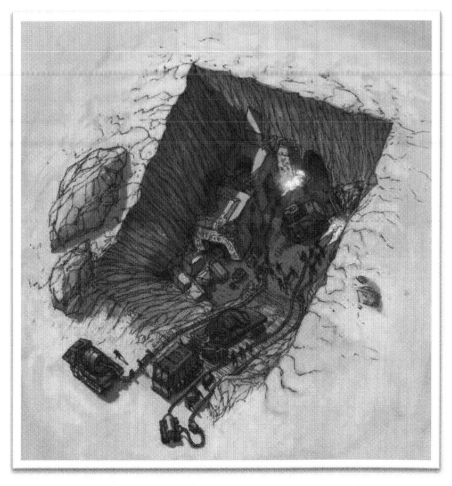

Figure 51. Drawing of excavation with ramp descending into the site. Courtesy of www.Gaia.com

They arrived here approximately 55,000 to 65,000 years ago and began to create hybrids of their species and the developing human population. [384]

Once again, if true, this information is ground-shaking across diverse fields of scientific study since it confirms that the elongated skulls found in places like Paracas, Peru belong to *another species of humans*, rather than being artificially created deformities.[385]

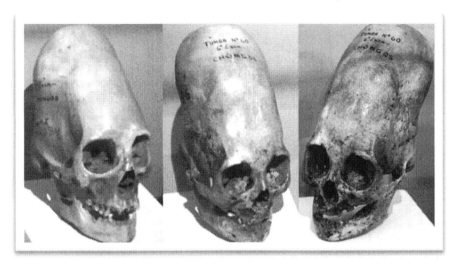

Figure 52. Skulls found near Paracas, Peru

What is critical to understand, according to Goode, is that many of the global elite view themselves as direct descendants of these pre-Adamite peoples, and consequently, view the Antarctica discovery as an event that corroborates their uniqueness, and fitness to rule.[386] He claims that the descendants of these Pre-Adamites occupy very senior positions in the Vatican hierarchy, where their identities are hidden by the elongated hats worn by Bishops and Cardinals. If Goode is correct, this would help explain the extraordinary influence the Vatican has held over a large segment of humanity for nearly two millennia.

It's important here to bring in the Thule Society and their belief that Aryans were descendants of the ancient Hyperboreans, who were giants with superior intellectual and psychic skills. It becomes easy to understand why such beliefs were shared by elites from many nations, thereby facilitating the eventual agreements that led to secret cooperation with the breakaway colony established in Antarctica by the Thule and other German secret societies.

In late 2016, Goode said that he has been told by a number of other sources independent of Sigmund that recent discoveries have indeed occurred in Antarctica:

> I have now had well over a dozen confirmations that indeed a HUGE F-ING discovery was made down in Antarctica. Indeed there are many, many types of ruins and artifacts strewn out across the continent/ group of islands and underground.... Many square miles of ruins have been detected w/only a small % that has actually been excavated. [387]

At that time, all Goode knew about the Antarctic discovery and excavations was through second-hand sources, but they were nevertheless consistent with other claims of a major discovery having occurred at Lake Vostok, and with the anomalies Brian claims he witnessed in Antarctica. Furthermore, the visit of VIP's to Antarctica in 2016 is powerful circumstantial evidence in support of the discovery of a buried civilization, part or much of which is located near Lake Vostok.

In early 2016, however, something truly extraordinary happened to Goode. He says he was taken down to Antarctica by an Inner Earth group, who gave him direct access to the buried Pre-Adamite civilization where he saw for himself the secret excavations that were occurring there.

CHAPTER 12

Corey Goode's Covert Reconnaissance Missions to Antarctica

Use of Cover Programs to Hide Classified Programs

Gathering reliable intelligence about current activities in Antarctica is very difficult due to the "need to know" aspect of the covert projects occurring deep under the ice shelves. Most public officials who travel to Antarctica are only given a highly sanitized version of projects occurring at the stations found on the surface. If visiting officials don't have a *need to know* what is happening deep inside the interior, they are only given routine tours of the surface facilities. Then they are shown open source science projects, which are effectively cover programs for the highly classified subterranean Antarctic projects.

In 2015, Edward Snowden leaked National Security Agency documents that showed how all classified programs have cover programs, along with contrived cover stories to maintain secrecy. The most highly classified programs are covered by less classified programs. This was illustrated by one of the documents released by Snowden showing how highly classified programs will be obscured behind less classified counterparts.

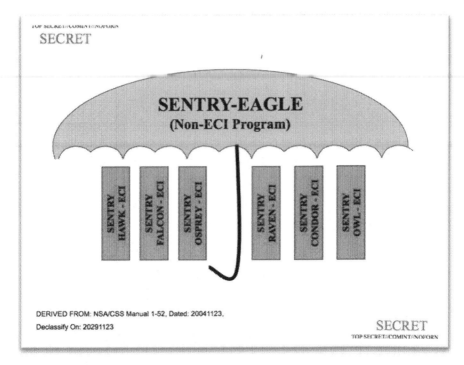

Figure 53. Leaked NSA document

Classified activities in Antarctica will similarly have to be covered up by a variety of cover programs. Public officials without a *need to know* are given a tour of the cover programs in Antarctica, which appear to be legitimate science projects covering topics such as weather monitoring, penguin research, environmental changes, atmospheric physics, etc. In this way, secrecy can be effectively maintained despite the presence of thousands of scientists and personnel in Antarctica witnessing anomalous events, while participating in a number of open source or unclassified projects.

Several public officials interviewed for this book said they did not see anything pertaining to a lost civilization or ancient artifacts during their visits to Antarctica. Both Congressman Nicholas Lampson (H.R. Texas, 1997-2005, 2007-2009) and Dr. Rita Coleman (head of the National Science Foundation) responded to a series of questions they were asked for this book, and said they did not witness or hear about such topics during a 1992 U.S. Congressional visit to Antarctica.[388] They each claimed they had only witnessed science projects, which are well known in open source literature such as the *Antarctic Sun*. Neither admitted to being briefed about classified projects.

Clearly, if there are highly classified U.S. projects happening deep under the ice shelf, or a German space program operating in unknown bases there, eyewitness confirmation has proved elusive thus far. The closest we have come is the testimony of a navy flight engineer, "Brian", who revealed that there is a large airplane-sized hole at the South Pole, which appears to be part of a highly classified program. The suppression of eyewitness testimony is evidenced by the debriefing Brian and his colleagues experienced after they flew over the hole during a rescue mission. They were instructed not to divulge what they had seen, and warned of dire consequences if they disclosed anything. To the present day, Brian refuses to divulge the other members of his team out of concern for the repercussions to them as well as himself.[389]

Congressman Lampson and Dr. Coleman visited Amundson-Scott South Pole station during their 1992 tour. When asked whether a large hole existed near the South Pole, each one answered that he/she had no knowledge of it. Yet, Brian's background and credentials lend authenticity to the accuracy of his report of what he personally witnessed back in the 1990's. Furthermore, Brian's observation of flying saucers over the Transantarctic Mountains on several occasions during flight missions suggests that a secret space program does exist in Antarctica. Significantly, this is a program that is not part of the oversight mechanism in place for the U.S. Antarctic Program, run by the National Science Foundation. The flying saucers Brian witnessed are evidence that an advanced aerospace program, possibly part of a German-led program, does operate in the Transantarctic mountain range. Finally, Brian's recollection of an incident involving missing scientists in Marie Byrd Land indicates that covert projects are occurring in Antarctica, and any scientists involved are debriefed with clear directives to maintain secrecy.

Based on Brian's testimony, we only have a sketchy understanding so far of covert programs in Antarctica and the continuing German presence there. The same is the case for a magnetic anomaly discovered near Lake Vostok, and Corey Goode's claims via information told to him by his USAF contact "Sigmund", and other insider sources, about a flash frozen buried civilization discovered under the ice mass.

However, it is Goode's recent remarkable eyewitness testimony that tells us the most about what is occurring in highly classified projects in Antarctica. To date, he claims he has been taken to Antarctica on two separate occasions where he witnessed advanced bases and excavation activities related to an ancient alien civilization that possessed advanced spacecraft. It is essential, therefore, to review and analyze these claims beginning with his 2016 trip where Goode contends he witnessed large industrialized bases connected to the German Antarctic space program (Dark Fleet), which extensively cooperates with a transnational corporate group; the Interplanetary Corporate Conglomerate (ICC).

2016 Trip to Secret Antarctic Bases

On May 14, 2016, Goode released an online "report" about Antarctica related events in bullet point form, comprised of information he recently acquired from his insider sources.[390] This report included mention of the Anshar, one of the seven Inner Earth civilizations that Goode says he has met with since 2015. In earlier online reports, he disclosed being taken to the main underground city belonging to the Anshar, where he witnessed their advanced technologies. Goode also previously described multiple encounters with Ka Aree, a High Priestess of the Anshar, who has acted as his guide and friend on many trips into the Earth's interior and deep space.

A month later, Goode gave me an informal "briefing" about recent events in Antarctica, including details about his first trip to the icebound continent onboard a spacecraft belonging to the Anshar to witness industrial facilities underground.[391] Goode's Antarctic reconnaissance tour took place between April 27–30, 2016. The facilities he witnessed were largely unknown to the Secret Space Program Alliance, formed in part largely by the US Navy's Solar Warden program, and therefore Goode acquired highly critical intelligence that he was asked to release as quickly as possible; so Goode briefed me to get the information out immediately. The next day, on June 16, 2017, I released a lengthy article entitled, "Secret Space Programs Battle over Antarctic Skies during Global Elite Exodus" on Exopolitics.org to summarize the data Goode shared with me.

The next month, in an interview on camera with *Cosmic Disclosure*, which was released on July 12, 2016, Goode described his initial visit to Antarctica in detail.[392] He began by describing being taken on an Anshar spacecraft possessing advanced viewing technology, for a reconnaissance mission that would give him relevant information for the Secret Space Program Alliance, which was beginning to question his value as an asset to them. Goode also mentioned that he was once again accompanied by Ka Aree:

> And she starts telling me, she said, "The technology of this craft is incredibly intuitive." She said, "Do not get lost in the data."... But she said this just before we started heading right towards a giant wall of ice, like an ice shelf. And we were heading directly towards it at speed....
>
> And we passed right through what had to be some sort of hologram. And soon as we punched

through, we saw an area carved out to where the largest ship that we have [an aircraft carrier] could go through and still have two or three times the height of the ship to the ceiling of this archway.

But there was all of this, I guess, steam. It was real foggy.

At the entrance, this like fog was coming out. And we headed directly into this basically ice tube. And we were flying all around, and there'd be areas ... all you'd see was ice, but then you would see, like, some little bit of rock outcroppings. And this was several miles that we went that way until everything started to then spread out and open up.[393]

What Goode has described appears to illustrate thermally heated ice caverns that are formed by steam coming from active volcanoes deep under the ice sheets. This is notable because later, Australian and New Zealand scientists confirmed in September 2017 that there are indeed networks of ice caverns deep under the ice shelf that are formed by thermal heat from active volcanoes.[394] The scientists pointed out that these could be heated to approximately 25 degrees Celsius (77 °F) and were very capable of hosting life.

Continuing, Goode explained how the Anshar craft overflew one of the hidden waterways deep under the Antarctic ice sheets. These waterways go through some of the large thermally heated ice caverns all the way out to the open sea:

And then we were flying over water. We're flying under ... over water, under the ice. And then we came up upon this little island area that had a little outpost – looked like an industrial city, but not a city. It was small.

And we're still going at a good clip. And all of a sudden, we're ... there's land below us. And you can't see it real well because it's very dark. And then we get to an area and you start seeing light, a lot of light, up ahead.

And the light was penetrating and refracting and reflecting out of this giant ice dome above the ice....

And there were these pools of water that had steam coming out of them. And it was obviously very warm. I could ... There were trees that looked about this big [Corey shows a distance of approximately 4 inches with his fingers.] that looked ... that had pine needles on them. But it was so dark, I couldn't make them out real well.

And we curve around this mountain ... and there were these mountain peaks that came up out of the ground and disappeared into the ice, the ceiling of the ice. It was bizarre. And this was obviously ... this ice pocket, or little igloo under the ice, was obviously created by the thermal activity. And this was in the northwest area of Antarctica. [395]

This is an accurate description of how the peaks or upper portions of mountains in Antarctic are covered over by the ice sheet, but surface levels may be exposed due to thermally created ice caverns at the base of the mountains. Scientists announced in August 2017 that the Northwest area of Antarctica is filled by active volcanoes (see Figure 61), thereby corroborating what Goode claims he saw during his April 2016 trip.[396]

Next, Goode described coming across the first major industrialized city he witnessed during his Antarctic reconnaissance mission:

We then came upon the first really big industrial city that I saw under there. And you could ... It looked like it was ... At one time, it was about this size, [Corey shows a distance of about 10 inches between his hands.] and then they had built it out [Corey spreads his hands as wide as he can.] over time....

Another thing, we saw all of these large triangle craft just ... I couldn't tell if they were hovering above the ground or parked on the ground because of the lighting conditions.

And it was pretty well lit from all of the industrial lights reflecting off the ice in the general area of where this complex was. [397]

Goode said he then saw large submarines that were able to access this underground city through hidden waterways from the coastline spanning well underneath the Antarctic ice sheets:

And as we flew over, we see two conventional-looking submarines and a few of those very large black subs. And they had these cranes on kind of like a train track that were positioned and unloading part of the almost egg-shaped black sub that the top kind of like slid back like that and exposed the inside. And they were unloading....[398]

In Goode's May 14, 2016 online report, he first mentioned what he had been told by 'Gonzales' about these huge submarines used to transport people and cargo to Antarctica:

Gonzales later confirmed that these people and supplies were in many cases being transported to Antarctica via "Black Submarines" that were "EM

Driven" and the "size of container ships". The water filled subterranean rift systems are so incredibly enormous that they have no trouble on their journey. Furthermore, the reports stated that the rift caverns had been modified into massive arched tunnels in ancient times.[399]

William Tompkins also said that the cavern system under the Antarctic ice was navigable by submarines, and the Germans had built enormous submarines for this purpose:

> They had access to these [Antarctic] caverns with their submarines. The Germans built massive *truck* submarines, which were enormous submarines, to transport all this stuff down to Antarctica. The submarines, still submerged, entered these Antarctica caverns through the underwater tunnels. The submarines went back through the tunnels, and they entered a lake where there were all these facilities, the cities, the naval bases, towns, and thousands of people.[400]

After describing the giant cargo ship sized submarines at the Antarctic industrial city, Goode offered more details about the craft carrying him and the Inner Earth crew:

> And it was at this point that I started wondering how many of these outposts are there? Is there anything under the ground? What ... You know, I started thinking all this stuff. And all of a sudden, all of these displays ... I mean, like holographic displays started popping up all around.
>
> And what I didn't mention is when we first entered this under ice area, the ceiling and the floors

turned transparent on this craft. [401]

On the holographic maps displayed on the wall panels of the Anshar vehicle, Goode was able to see the locations of the industrial areas in Antarctica as they appeared:

> So all these displays started popping up, and I started looking at it and I was like, "Wait, I need to see, you know, see what I'm supposed to." .... And I really wasn't understanding all I was seeing — everything that was popping up. I saw sort of the map of the area we were in, and I saw two large areas and then four smaller outpost areas on the map. [402]

A map has been provided by Goode with the six industrial areas he viewed circled to show their approximate size and location.

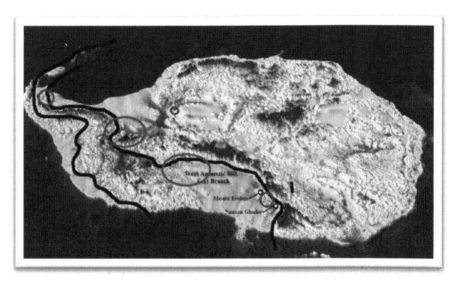

Figure 54. Map of Antarctica with circles designating locations and relative sizes of industrial complexes. Courtesy of Sphere Being Alliance.

Significantly, the locations are all near subterranean waterways and in a region of Antarctica where large numbers of active volcanoes exist. Therefore, Goode's account of submarine accessibility along with thermal activity taking place at the underground facilities is consistent with scientific data about these specific underground regions of Antarctica. The six Antarctica bases belonged to one of the secret space programs, according to Goode: the Interplanetary Corporate Conglomerate.

Goode went on to describe the next unusual site they came upon:

> We pass over that industrial city. It gets darker again. And then all of a sudden, we're over like a bay area. And across the end of the bay area, we could see what looked ... which was another city that showed up on the map that popped up for me.
>
> And I thought we were heading right towards it, but we stopped over the water. And I look up, and I see this huge ... The ice was 200 - 300 feet above the surface of the ground in most of the area, but in this bay area, it was 600 feet or 500 feet. It was much higher.
>
> And there was this large hole, you know, 70 - 90 meters wide, that went up and then went off at an angle above us. It was a huge ... and it's apparently where all of the thermal heat is escaping ... So we stopped underneath it. And I think we're about to shoot up through it. But instead, "phst", we go into the water, like flat, and like a belly flop almost, but just "phew" into the water. And we're underwater. And it's dark. It's pitch black. [403]

What's of special interest here is that the hole Goode reported seeing is similar to the one Navy engineer "Brian" claims he witnessed near the South Pole. Goode stated the hole going up was 70-90 meters wide, which is comparable to Brian's estimate that a C-130 cargo plane could easily fly into the hole he saw. Since Brian was a flight engineer during his Navy service with the Antarctic Development Squadron Six, his estimate can be deemed accurate.

This indicates that the South Pole hole was created by thermal energies emanating from the interior, which formed a cavern in the ice that grew upward, over a distance of one to two miles all the way to the surface. It's very possible that there are many thermally created holes stretching vast distances up through the layers of ice to the Antarctic surface.

Goode then detailed how the Anshar vehicle traveled through a subterranean river, which was accessible to submarines traveling the expanse from the coastline:

> And you know, we're, you know, traveling under the water. And then all of a sudden, we enter into this cavern type area where obviously these subs are coming and going from ... I don't know how far, we head back a ways. And I see two or three of those subs coming with a long line of tiny little bubbles behind it and a light out the front. And this was obviously a rift.

> And the rift started to curve around, and we traveled around curving, and then we saw this huge arch that was, I mean, I don't ... It was so big. And it ... what it had done is, a rift was curving around this way, and this was an arch that connected two rift or tunnel areas. And it was ancient. [404]

Goode has provided an artistic illustration of the tunnel system and arch through which submarines can travel into and from

Antarctica. What's important here is that Goode is referring to a tunnel system that extends all the way from Antarctica to South America, where another tunnel system stretches all the way up the continent, even into North America.

Figure 55. Illustration of underwater tunnels described by Goode running into South America. Courtesy of www.Gaia.com

William Tompkins spoke of the existence of a similar tunnel system reaching across the African continent, and connecting into Europe and beyond. During WWII, according to Tompkins, Nazi Germany began moving equipment and personnel through this tunnel system across Europe into Africa, and then all the way to Antarctica. He said that as much as half of the personnel and resources which found their way to Antarctica came through this ancient tunnel system linking Europe, Africa and Antarctica. Tompkins also mentioned the tunnel system in an interview and told how submarines could use these for access to the underground Antarctic bases.[405]

It is known with certainty that Nazi Germany commissioned a thorough study of natural caverns systems in Germany and Occupied Europe for the construction of underground manufacturing centers. Therefore, it's quite possible that during the development of these natural caverns in Europe, the Germans stumbled onto the continent's natural cavern system extending all the way down to Antarctica. Certainly, the Nazi's Reptilian allies would have helped them find and use this cavern system, in order to escape the collapse of the Third Reich. Goode's observation of another tunnel system extending through the Americas also makes Tompkins' claims more plausible.

What Goode witnessed during his April 2016 Antarctic visit was time sensitive information, since it offered new intelligence for the SSP Alliance to aid their ongoing efforts to counter or neutralize some of the activities led by the Dark Fleet and Interplanetary Corporate Conglomerate. Goode confirmed the SSP Alliance found the information very useful for their ongoing operations against the Cabal, some of which involved direct military confrontations. During this period, Goode was also releasing information given to him by the SSP Alliance. Before discussing Goode's second trip to Antarctica, it's first worth examining the briefings he says he received about an aerial confrontation that took place over Antarctica in early 2016.

Battle over Antarctica

According to Goode, briefings he received about what was happening there around the time of his April 2016 visit came from two different sources. One source, "Gonzales" (a pseudonym), was his primary liaison with the SSP Alliance. Gonzales, an alleged

US Navy Lieutenant Commander, had also acted as Goode's initial contact with the SSP Alliance, which is comprised of the Navy's Solar Warden Program and defectors from other secret space programs.

Goode's second briefing source was 'Sigmund', an alleged senior officer with a US Air Force run secret space program, which is comprised of USAF Space Command, the National Reconnaissance Office (NRO), National Security Agency (NSA) and the Defense Intelligence Agency (DIA). As mentioned earlier, Sigmund was involved in abducting and interrogating Goode during much of 2016 because the accuracy of his disclosures had startled leaders within the USAF run SSP and raised many questions. Goode refers to the USAF program as a "lower level SSP" since the advanced aerospace technologies it uses are hand-me-downs from other more advanced space programs run by the Navy (Solar Warden) and the Interplanetary Corporate Conglomerate.[406]

Goode's most important briefings during this time concerned an alleged air war over Antarctica that occurred in early 2016. Goode first wrote about events leading up to this confrontation in his May 14, 2016 online "report", in which he described what 'Gonzales' and 'Sigmund' had told him of an exodus occurring, involving elite groups fleeing to South America and Antarctica.

> Reports came in for approximately 6 months that high level syndicate groups were moving huge amounts of personal items and supplies to South American underground bases most noted in Brazil. More recent reports stated actual family members and high-ranking syndicate members were pouring into these underground bases like ants before a storm.[407]

Goode elaborated on these events when he gave me the

"briefing" in June, and explained that syndicate groups (global elites/Cabal/Illuminati) feared huge solar storms predicted to hit the Earth.[408] A so-called "solar kill-shot" had long been predicted by the remote viewer, Ed Dames, who stated in an interview on March 21, 2016 that it was now "imminent".[409]

Goode also clarified that syndicate groups could not easily leave the Earth for refuge on off-planet locations due to a recent airspace lockdown both "on and around" the Earth by "the lower level SSP's".[410] The latter had deployed technology establishing an "Earth Defense Grid" coordinated from an "air (and space) traffic control" established on the Moon, called Lunar Operations Command (LOC).[411] Lunar Operations Command, currently controlled by the Interplanetary Corporate Conglomerate, was coordinating with the lower level SSP's. With the Earth Defense Grid able to stop spaceship exists and entries, the cabal/syndicate groups chose to flee instead to Antarctica/South America where underground locations provided some safety. Goode stated that during this time, the lower level SSP's were also instructed not to clear SSP Alliance flights into Earth's airspace, which was a temporary mandate.

Coincidently, President Obama visited Bariloche, Argentina on March 24, 2016, which had become the notorious unofficial headquarters of the "Fourth Reich" once Adolf Hitler fled there after World War II.

In February 1960, President Eisenhower traveled to Bariloche where he negotiated the "Joint Declaration of Bariloche" with the Argentinian President concerning Peace and Freedom in the Americas.[412] However, the real topic of negotiations most likely concerned the deals that were first reached in 1955, which would be altered to further place the U.S. military-industrial complex firmly under the control of the Fourth Reich. Here, it's worth remembering the CIA document shown at the end of chapter 3, which is evidence that Hitler moved from Colombia back to Argentina in February 1955. Given the additional documentary research confirming that Hitler had settled in Bariloche, it's almost certain that during Eisenhower's visit, he met with Adolf Hitler who likely held an honorific position with the Fourth Reich.[413] Furthermore, it's very possible that Bariloche has easy access to the underground cavern system that extends to Antarctica, which submarines can use as Goode witnessed during his 2016 visit. If so, then the large lake near Bariloche may well offer a means for submarines to travel back and forth to Antarctica.

The alliance between the Germans in Antarctica and the U.S. military-industrial complex not only led to the emergence of the Interplanetary Corporate Conglomerate (ICC), but also to placing in charge of the German bases in Antarctica. Therefore, it is highly likely that President Obama's visit to Bariloche was to finalize new deals with the ICC/Germans to facilitate their desire to move a large number of people and cargo to safe locations in South America and Antarctica for their exodus.

In his May 14, 2016 online report, Goode described a battle that had taken place over Antarctica as a consequence of the personnel and resources covertly transferred to the secret industrial facilities hidden there:

> One of the most interesting things that came out of this briefing [by Gonzales now safely located at a secret Kuiper Belt base] was that there had recently been reports of 6 large cruisers (teardrop shaped) were in the process of leaving the atmosphere after breaking the surface of the ocean near the coast of Antarctica. Dozens of "Unknown Chevron Craft" swarmed these cruisers and attacked the leading two craft causing massive and shocking damage. The cruisers broke off their attempts to leave orbit returning to below the surface of the ocean where they came from.[414]

Goode has provided an illustration of the battle that took place.

He said that the teardrop-shaped cruisers belonged to syndicate groups aligned with the "Dark Fleet" (a space program first established in Antarctica by Nazi Germany/German Secret Societies), which later allied itself with the U.S. military-industrial complex.

Figure 56. Battle over Antarctica. Courtesy of www.Gaia.com

Goode added that the SSP Alliance was not sure who the chevron-shaped spacecraft belonged to, but speculation linked them to the "Earth Alliance", a consortium of "White Hats" from various national militaries working closely with the BRICS nations (see Figure 1 for different SSP alliances).[415]

According to Goode, the Antarctic battle was not an isolated incident:

There has been a major uptick in conflicts just

outside and within our atmosphere between craft of various groups that have involved the shoot down of a number of craft … Gonzales reported dozens of underground/ocean conflicts that have involved the use of exotic weapons as well as an uptick in the use of weather modification weapons by both the various syndicates and elements of the Earth Alliance.[416]

Events in Antarctica were being monitored closely by different nations and/or space programs, and according to Goode, huge spherical craft over Antarctica appeared to also be conducting surveillance operations:

Reports came in for approximately 6 weeks detailing "huge spherical craft" in geostationary orbit above the continent of Antarctica. These reports came from 5 different sources and described the spheres as being huge, metallic, shiny with one row of portholes going around the sphere. One speculated that these craft were of Russian origin. [417]

Goode then explained that the craft were operating over large areas of the Southern Hemisphere, including Australia.

It's possible that the craft are related to the "Cosmospheres" allegedly developed by the Soviet Union, which are described at length in the Peter Beter audio files.[418] Notably, Beter held the position of General Counsel for the Export-Import Bank of the United States (1961-67), and had high-level sources who he claims confided what was happening to him behind the scenes in space up to the early 1980's. The Cosmospheres, if Soviet, established their weapons dominance in near Earth orbit, where they had military skirmishes with craft belonging to the USAF Space Command and NRO. According to Goode, these operate about 400 miles above the Earth, and occasionally send their most advanced craft to the Moon.

It is, therefore, very likely that the large spherical objects observing the Antarctic space battles were indeed Cosmospheres now under the control of President Putin and the Russian Federation. It is feasible that they provided intelligence used by the chevron-shaped spacecraft that intercepted and turned back the larger teardrop-shaped craft, leaving with their global elite passengers.

**Figure 57. Illustration: Spherical UFOs may be Russian Cosmospheres.
Courtesy of www.Gaia.com**

Corey Goode's Second Trip to Antarctica

In early January 2017, Goode said he was again taken by the Anshar to Antarctica on another reconnaissance mission. This time he witnessed the first scientific excavations of ruins from an ancient "flash frozen" civilization buried under two miles of ice. While the discovery dates back to the first Nazi German expedition in 1939 according to Goode, it is only since 2002 that excavations by archeologists and other scientists have been allowed.[419] The archeologists have allegedly prepared documentary films and academic papers on the ruins, and their eventual release will astound the scientific community.

Coincidently, this corresponds with the initial unconfirmed reports, investigated by Richard Hoagland and Mike Bara, that an ancient city was discovered near Lake Vostok around 2002. Goode's January date also coincides with the contrived news release concerning an @tlantis TV crew that went missing after filming an underground ancient city. Goode's extraordinary information casts light on these puzzling 2002 developments, which carry the hallmark of an effort by the intelligence community to hide the truth in plain sight. This means that the core story is real, but many of the details surrounding the discovery and secret filming, in this case, are contrived to throw serious investigators like Hoagland and Bara off the trail.

Goode was told that three oval-shaped craft were discovered at the site, revealing that the Pre-Adamites were extraterrestrial in origin, and had arrived on Earth about 55,000 years ago: "there were three [buried spacecraft] that were extremely large. They were motherships."[420] He further clarified "that there was one miles-long craft that was up to three miles and oval-shaped, and then there were two smaller, I guess, support-type craft."[421]

The largest of the three spacecraft had been excavated and found to have many smaller craft inside. These smaller craft were test flown, according to Goode, inside the larger mothership by government operatives. Most surprising, the mothership was found to also contain Pre-Adamites:

> And in the largest craft, there have indeed been located a bunch of beings in stasis, and they are the original beings from, I guess, Mars, that had come here – the original Pre-Adamites. [422]

The Pre-Adamite civilization, at least that portion of it based in Antarctica, had been flash frozen in a cataclysmic event that had occurred roughly 12,000 years ago.

Corroboration for Goode's claim of three extraterrestrial craft having been discovered under the Antarctic ice comes from Dr. Pete Peterson who claims he worked on multiple classified government projects. Peterson first became known to the public when he appeared in three interviews on *Project Camelot* in 2009 where he discussed his involvement in various classified projects.[423] In his most recent set of interviews on Gaia TV's *Cosmic Disclosure*, Peterson discussed his first-hand knowledge of three extraterrestrial ships that were buried deep under the Antarctic ice sheets. He stated:

> [T]here are three layers of these things ... Three crashes. And strangely one's down about a mile, and one is about two miles, and one's about three miles.... There were things that were similar, which told me that there was probably commercial traffic/communication between those societies.... And wholly different controls. Those people had three fingers, so you had a place that your hand would fit into an indentation in a control surface. And it was ... You had two thumbs, opposing thumbs, one on either side, and a main finger.[424]

While Peterson described crashed spacecraft found in three different layers separated by a mile of ice, Goode had been told the three craft were in close proximity to one another. Nevertheless, it is significant that both Goode and Peterson were told about the discovery of three large extraterrestrial motherships buried under Antarctica, which were used to develop one or more civilizations.

Goode has also been told by his contacts that the most advanced technologies, and the remains of Pre-Adamites themselves, have been removed from one archeological site that will be made public. Teams of archeologists have been working with what remains, and have been told to maintain secrecy over other things they have witnessed.[425] In addition, select ancient artifacts from other locations will reportedly be brought in from vast top secret warehouses and seeded into the archeological site slated for public release. In the impending announcement about the Antarctica excavations, emphasis will be placed on the terrestrial elements of the flash frozen civilization in order not to overly shock the general population.

Up until early January 2017, everything Goode knew about the Antarctica excavations had come second-hand from his insider sources. That changed when Goode was taken to Antarctica to witness first-hand the ruins and the excavations underway. I met with Goode on January 24, 2017 in Boulder, Colorado, where he gave me a personal "briefing" about his second Antarctic trip.[426] The next day, I released an article summarizing important details of his mission entitled, "Visit to Antarctica Confirms Discovery of Flash Frozen Alien Civilization" on Exopolitics.org. Ka Aree led the reconnaissance mission aboard the Anshar spacecraft, which Goode learned was being conducted for his benefit. Another key figure on the mission was "Gonzales". Goode had exposed Gonzales during his involuntary abductions and interrogations by "Sigmund" in 2016, and Gonzales subsequently became a liaison between a Mayan Secret Space Program and the SSP Alliance, an assignment which no longer required his presence on Earth (see Figure 1 for Mayan and other SSP's).

Two other Inner Earth civilization representatives also were present for the reconnaissance mission. Goode and the others were taken by the Anshar spacecraft to an unexcavated portion of the ruins. This was an area that the nearby scientific teams had not yet reached, so it was still pristine and showed the full extent of a civilization that had been devastatingly flash frozen. Goode described seeing bodies twisted and contorted in various frozen states indicating the catastrophe had clearly been unanticipated. The scene was similar to ancient Pompei which was buried under tons of volcanic ash and lava, but here the Pre-Adamite civilization was buried under tons of snow and ice.

Figure 58. Bodies found during excavation of Ancient Pompeii

Goode noted that the Pre-Adamites were very thin. Examination of their bodies made it evident that they had evolved on a planet with a much lower gravitational environment. In addition to the Pre-Adamites, Goode also saw many different types of normal-sized humans, some of whom had short tails, while others displayed elongated skulls similar to the Pre-Adamites. The conclusion Goode drew was that the Pre-Adamites had conducted biological experiments on the indigenous humans of the planet.

Gonzales had an instrument for taking biological samples which he jabbed into various frozen bodies. He also carried a camera and took many photos. The biological material and photos would be given to SSP Alliance scientists for study. Additionally, there were scrolls made of a metallic alloy, which were rolled up and displayed some kind of writing on them. The Anshar and the other Inner Earth representatives collected as many of these scrolls as was possible.

In earlier reports Goode released publicly, he described the Anshar Library as being remarkably extensive and possessing many ancient artifacts from multiple civilizations.[427] The Anshar were apparently adding the historical records of this flash frozen civilization to their library. During the team's time at the pristine site, Goode noted that his party was not seen by the scientists and archeologists working on excavations in another part of the vast ruins. The Anshar ship had traveled directly through the ice to get to the ruins, and Goode recalled how the ship could easily move through walls using advanced technologies.

The importance of Goode's January 2017 trip to Antarctica is that it confirms his earlier intelligence from various sources, including the accuracy of it by the USAF officer, Sigmund.[428] The Antarctica excavations proved to be quite real and Goode was now an eyewitness to it. Goode's visit and his verification of the Antarctica discovery is highly valuable. It is also brings a disturbing confirmation of Charles Hapgood's theory that pole shifts have been a regular occurrence in Earth's history.[429] The flash frozen Pre-Adamite civilization is not the only case of this type of catastrophe impacting an ancient civilization. Could it happen again today?

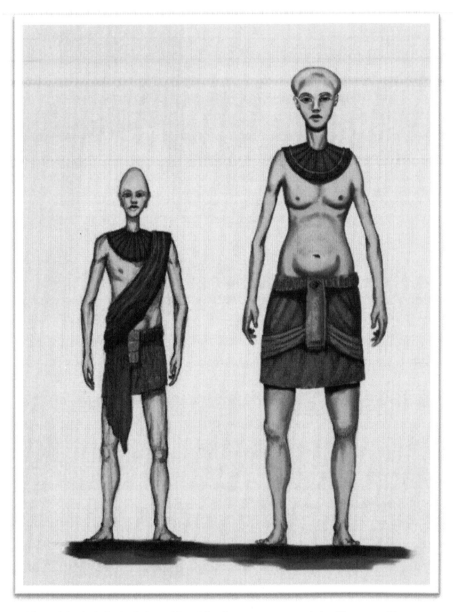

Figure 59. Artistic depiction of Pre-Adamite next to normal sized human with cone shaped head. Courtesy of www.Gaia.com

Will a Pole Shift Happen if Antarctica's Volcanoes Become Active?

Scientists studying Antarctica's newly discovered volcanoes have observed a worrying global pattern that may repeat itself in Antarctica.

> The most volcanism that is going in the world at present is in regions that have only recently lost their glacier covering – after the end of the last ice age. These places include Iceland and Alaska.
>
> Theory suggests that this is occurring because, without ice sheets on top of them, there is a release of pressure on the regions' volcanoes and they become more active.
>
> And this could happen in west Antarctica, where significant warming in the region caused by climate change has begun to affect its ice sheets. If they are reduced significantly, this could release pressure on the volcanoes that lie below and lead to eruptions that could further destabilize the ice sheets and enhance sea level rises that are already affecting our oceans.[430]

Scientific data indicates that this heating process has been underway in Antarctica for decades. A fifty year study from 1956 to 2006 showed that the west Antarctic region is heating up faster than east Antarctica. A map showing the temperature difference per decade was released in 2009 by NASA (see Figure 61).[431] It vividly shows how an unknown geological process is heating up west Antarctica more rapidly than east Antarctica.

As to what geological process is responsible for the heating up of west Antarctica, we can find an answer in recent scientific studies. Maps of temperature increase in west Antarctica correspond closely to the region occupied by the newly discovered volcanoes. A comparison of the scientific data displayed in the two maps respectively show temperature increase and newly discovered volcanoes, which points to a momentous conclusion – west Antarctica's ice sheets are being melted due to steadily increasing volcanic activity under the ice.

The melting of the West Antarctica ice sheets supports Goode's claims that frozen artifacts buried there for millennia are now being exposed by this process, and governments are scrambling to secretly send excavation teams to study these newly exposed finds. Indeed, if the melting continues, then it is very possible that Antarctica's volcanoes will be the colossal engine forcing a disclosure of ancient civilizations, advanced technologies and/or extraterrestrial life.

This possibility was first mentioned by whistleblower Emery Smith, who worked for the U.S. Air Force as a medical technician before being recruited into covert operations. In a January 2018 interview on *Cosmic Disclosure*, Smith described what he had learned about Antarctica's artifacts being exposed by melting ice:

> You know, it won't be the people. It won't be us exposing this. It's going to be Earth that exposes it because of the warmth. They can't fight the heat right now. And since they can't fight the heat, it's going to be really hard to explain when some of the snow melts in the next year, and this giant thing starts being exposed and different metals.... So Gaia will be Disclosure. Earth will be the one who discloses it, which is beautiful.[432]

While the melting of the Antarctic ice sheets can force the disclosure of many frozen secrets long buried across the

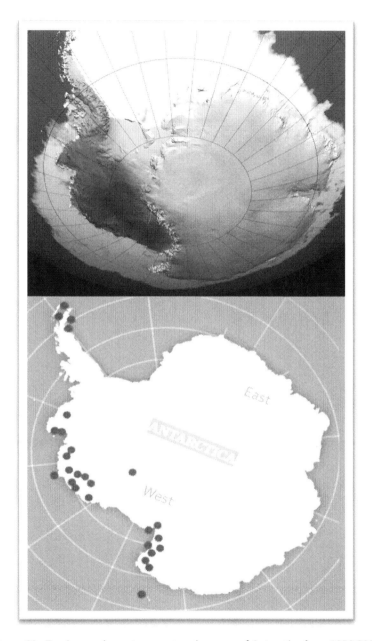

Figure 60. Top image shows temperature increase of Antarctica from 1956-2006 (Credit: NASA/Goddard Space Flight Center Scientific Visualization Studio); bottom image shows 2017 discovery of 90 new volcanoes in Antarctica.

inhospitable continent, there are also major global problems the melting will create. The most well studied is the rapid rise in sea levels as the ice sheets melt, which can result in the disappearance of many coastal regions around the planet.[433] There is, however, an even more disturbing geological event that may lie ahead. As the west Antarctic ice sheets melt, the weight distribution around the geographic South Pole will dramatically alter. This may even trigger the cataclysmic geological event that Hapgood warned the world about – a Pole Shift!

How a Pole Shift could occur was summarized by Albert Einstein in his foreword to Hapgood's 1958 book, *The Earth's Shifting Crust*. Einstein described how the weight distribution of ice in the polar regions directly impacted the centrifugal momentum of the Earth responsible for the axis of rotation. He explained how changes in weight caused by a build-up of ice, which would also include the converse phenomenon of a melt-off, could result in a Pole Shift:

> In a polar region there is continual deposition of ice, which is not symmetrically distributed about the pole. The earth's rotation acts on these unsymmetrically deposited masses, and produces centrifugal momentum that is transmitted to the rigid crust of the earth. The constantly increasing centrifugal momentum produced in this way will, when it has reached a certain point, produce a movement of the earth's crust over the rest of the earth's body, and this will displace the polar regions toward the equator.[434]

Put simply, as west Antarctica melts, the new weight distribution at the South Pole could change the Earth's axis of rotation.

Conclusion

Goode's two visits to Antarctic in 2016 and 2017, along with Peterson's assignments there, provide the only eyewitness accounts of the extensive covert projects occurring deep under the Antarctic ice sheets. Their testimonies are corroborated by scientific discoveries in 2017 concerning an extensive network of thermally heated ice caverns created by Antarctica's submerged or under ice volcanoes.[435]

Goode's witnessing of the large industrial bases owned by the Interplanetary Corporate Conglomerate and the Dark Fleet helps corroborate evidence presented in this book; German secret society members and Nazis did indeed escape to Antarctica and built large bases in caverns hidden under a mile or more of ice.

Goode's revelations also support the existence of a vast network of subterranean rivers under Antarctica's ice sheets which are fully navigable by submarines, and link up with a subterranean tunnel network extending deep into South America, and as claimed by Bill Tompkins, possibly into Africa/Europe. Finally, Goode's testimony concerning the discovery of the Pre-Adamites in suspension chambers found in spacecraft buried deep under Antarctica's glacial surface raises many intriguing questions about the relationship between the global elite (cabal) and the Pre-Adamites, with a possible connection between the Pre-Adamites and the Fallen Angels described in the apocryphal "Book of Enoch".

CHAPTER 13

Antarctica's Secret History as
an Extraterrestrial Refugee Colony

Galactic History: The Mars - Antarctica Connection

Soon after his second Antarctic visit, Corey Goode revealed more startling information about Antarctica's ancient past as an extraterrestrial refugee colony, established roughly 60,000 years ago. In a *Cosmic Disclosure* TV episode aired on February 21, 2017, Goode described what he witnessed first-hand during that second excursion. He also relayed details about the extraterrestrial's history, which he had read about during his alleged service in a "20 and back" program with Solar Warden.[436] He said that the "Pre-Adamites" were originally from Mars and a Super Earth (several times Earth's diameter), whose remains now form the asteroid belt.

During the Pre-Adamite's long history, when Mars was still a moon of the Super Earth, Goode explained, the inhabitants of both Mars and the Super Earth fought a series of hi-tech wars. Approximately 500,000 years ago, these wars came to a climactic end with the Super Earth's obliteration. At the time, Mars had abundant water and an oxygen-rich atmosphere to host a large population on its surface. Intriguingly, back in July 2013, scientists confirmed the existence of abundant water and oxygen on Mars in its distant past.[437]

According to the records Goode accessed, the events that led to the Super Earth's destruction also wiped out much of the surface population on Mars, and removed the bulk of its atmosphere:

> It was postulated that Mars was most likely a moon of that Super Earth, and that it was damaged heavily on one side by massive impacts. And that most likely also stripped away its main atmosphere at the time, and it never recovered.[438]

The force of the Super Earth's destruction propelled Mars into its present planetary orbit. So, was there really a Super Earth destroyed in a titanic battle, which resulted in one of its moons being thrust into its present orbit, as the planet Mars?

The first major scientist to seriously research the possibility that the asteroid belt is from the remains of a former planet was Dr. Thomas Van Flandern, the chief astronomer at the U.S. Naval Observatory. He wrote several academic papers on what he described as the "Exploded Planet Hypothesis", which examined how our solar system's main asteroid belt arose from the destruction of a large planet that Mars once orbited:

> Putting all this evidence together, we have strong hints for two original planets near what is now the main asteroid belt: hypothetical "Planet V" and "Planet K". These were probably gas giant planets with moons of significant size, such as Mars, before they exploded ... The above summarizes evidence that Mars was not an original planet, but rather a moon of a now-exploded planet occupying that approximate orbit. Many of these points are the expected consequences of having a massive planet

blow up nearby, thereby blasting the facing hemisphere and leaving the shielded hemisphere relatively unscathed. Especially significant in this regard is the fact that half of Mars is saturated with craters, and half is only sparsely cratered.[439]

Van Flandern's "Exploded Planet Hypothesis" lends scientific support to Goode's claims about Mars being a former moon of a giant planet (Super Earth) that was destroyed long ago.

An enormous amount of debris from the Super Earth, as pointed out by Van Flandern, hit Mars with incredible force. This meant that the latter's surface cities on one side of the planet were totally destroyed, and most of its atmosphere was lost. This made life on Mars' surface very precarious at best, and according to Goode, led to planetary evacuation by the Martian survivors. What lends further plausibility to this scenario is a declassified Central Intelligence Agency (CIA) document, which was released on August 8, 2000.[440]

The declassified document reveals that in 1984, the CIA employed a psychic "remote viewer" to look at a region of Mars as it was approximately one million years ago. The remote viewer, who was not aware that the coordinates given to him were on the planet Mars, described seeing pyramids, futuristic technologies and a very tall human-looking species facing impending environmental calamity. What makes the CIA document remarkable is that the coordinates provided to the remote viewer were of the Cydonia region. This specific area had been photographed by the Viking Orbiter in 1976, heralding the first detailed images of Mars. Cydonia became famous after a succession of researchers claimed that the images of this region displayed a face, ruins of a city, and pyramids.

The first reference to artificial structures found in Cydonia dates to an October 25, 1977 *National Enquirer* article titled, "Did NASA Photograph Ruins of an Ancient City on Mars?"[441] It's worth

Figure 61. Viking image of Cydonia with Google Maps coordinates for Mars.

pointing out that the *National Enquirer* was a tabloid run by Gene Pope, a CIA asset who was trained in "psychological warfare".[442] Overseen by Pope, the *National Enquirer's* chief purpose was to hide the truth in plain sight by releasing it in sensationalized news stories with questionable sources that would be widely ridiculed by the general public. Subsequently, any academics or scientists prepared to investigate such sensational claims faced derision from their peers and ruining their professional careers.

Nevertheless, competent researchers did turn their attention to the now controversial Viking Orbiter images and found that they indeed did appear to show an artificially created "Face on Mars", nearby ruins dubbed "Inca City", and even pyramids. The first objective analysis of the Viking data was published in 1982 in *Omni Magazine* by researchers Vincent DiPietro, an electrical engineer, and Gregory Molenaar, a computer engineer. Their 1982 Omni article was an extract from their 77-page book, *Unusual Martian Surface Features*, also released that year. They were soon followed by other independent researchers including Richard Hoagland, who in 1987 authored, *The Monuments of Mars: A City on the Edge of Forever.*[443]

What this short review of the history of the Viking images of Cydonia tells us is that while researchers such as DiPietro, Molenaar and Hoagland were widely ridiculed by their scientific peers for their analyses and conclusions, the CIA was paying very close attention. Declassified CIA documents confirm that remote viewing was taken very seriously by the Agency and other intelligence services. Significant funding went into studying the usefulness of remote viewing as an "intelligence gathering tool".

The conclusion was that remote viewing had sufficient accuracy to be utilized for fieldwork as the following declassified CIA document, dated May 9, 1984, clearly affirms:

a. Remote Viewing (RV) is real, it is accurate, is replicable, is being pursued by at least the CIA, Navy, Army and Pentagon in general. That it is being pursued for intelligence and military applications.

b. The government's interest in RV is clearly "applications" oriented.

c. Stanford Research Institute International (SRI-I) is the key institute involved with the government in RV research and Development. The specific individual is Dr. Hal Puthoff (see Figure 63).

Among the most accurate of remote viewers described in the CIA documents was the famed psychic, Ingo Swann. In his 1998 book, *Penetration,* Swann described how the CIA utilized his

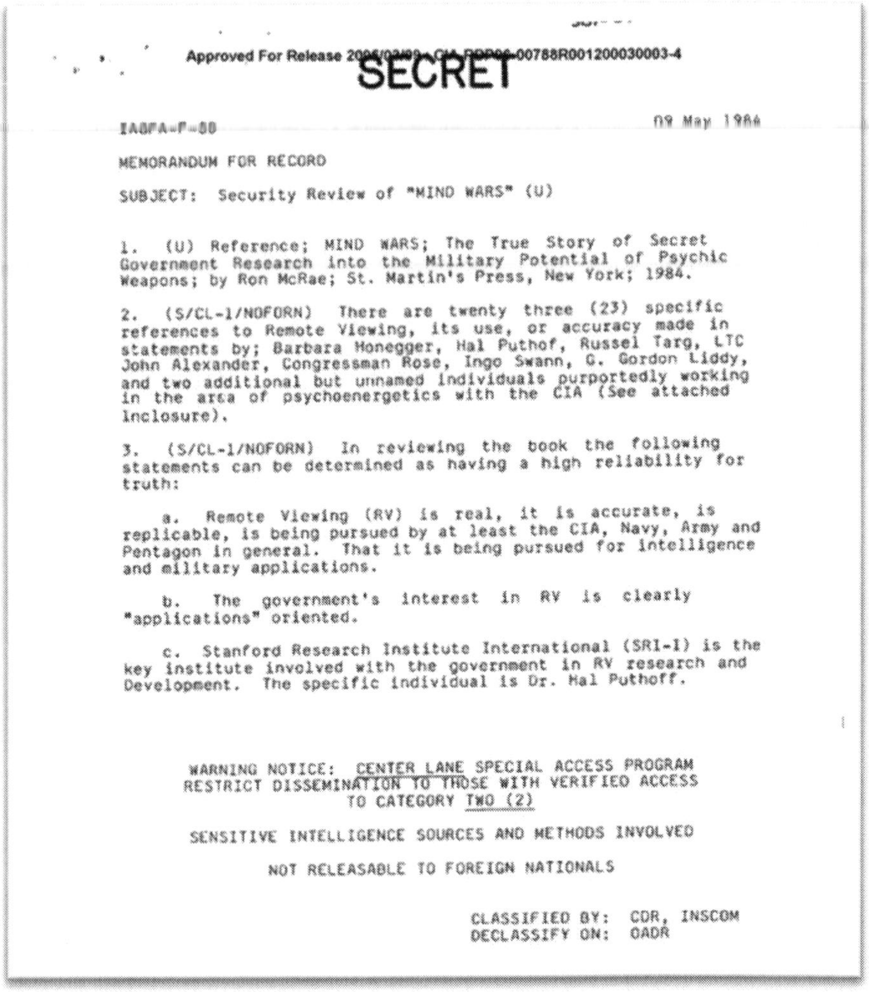

Approved For Release 2006/03/09 CIA-RDP96-00788R001200030003-4

SECRET

IAOFA-F-88   09 May 1984

MEMORANDUM FOR RECORD

SUBJECT: Security Review of "MIND WARS" (U)

1. (U) Reference; MIND WARS; The True Story of Secret Government Research into the Military Potential of Psychic Weapons; by Ron McRae; St. Martin's Press, New York; 1984.

2. (S/CL-1/NOFORN) There are twenty three (23) specific references to Remote Viewing, its use, or accuracy made in statements by; Barbara Honegger, Hal Puthof, Russel Targ, LTC John Alexander, Congressman Rose, Ingo Swann, G. Gordon Liddy, and two additional but unnamed individuals purportedly working in the area of psychoenergetics with the CIA (See attached inclosure).

3. (S/CL-1/NOFORN) In reviewing the book the following statements can be determined as having a high reliability for truth:

     a. Remote Viewing (RV) is real, it is accurate, is replicable, is being pursued by at least the CIA, Navy, Army and Pentagon in general. That it is being pursued for intelligence and military applications.

     b. The government's interest in RV is clearly "applications" oriented.

     c. Stanford Research Institute International (SRI-I) is the key institute involved with the government in RV research and Development. The specific individual is Dr. Hal Puthoff.

          WARNING NOTICE: CENTER LANE SPECIAL ACCESS PROGRAM
          RESTRICT DISSEMINATION TO THOSE WITH VERIFIED ACCESS
                         TO CATEGORY TWO (2)

          SENSITIVE INTELLIGENCE SOURCES AND METHODS INVOLVED

               NOT RELEASABLE TO FOREIGN NATIONALS

                              CLASSIFIED BY: CDR, INSCOM
                              DECLASSIFY ON: OADR

Figure 62. Declassified CIA Document supporting Remote Viewing

remote viewing skills at length.[444] One assignment in 1975 was to spy on secret bases on the Moon. In chapter five of his book, Swann opens by describing his remote viewing of the Moon for a CIA officer Axelrod/Axel:

> Back at work, Axel gave me Moon coordinates, each set representing specific locations on the

Moon's surface. At some of the locations there seemed to be nothing to see except Moonscapes. But at other locations? – well there were confusions, and I perceived a lot that I could not understand at all. I made a lot of sketches, identifying them as this or that, or looking like something else. Without comments, Axelrod quickly took possession of each sketch, and I was never to see them again.

I found towers, machinery, lights of different colors, strange-looking "buildings." I found bridges whose function I couldn't figure out. One of them just arched out – and never landed anywhere. There were a lot of domes of various sizes, round things, things like small saucers with windows. These were stored next to crater sides, sometimes in caves, sometimes in what looked like airfield hangers. I had problems estimating sizes. But some of the things were very large.[445]

Given the off-planet locations assigned to top remote viewers like Swann, it is not surprising that the CIA was later interested in what a remote viewer could reveal about the origins of the pyramids and other artificial Mars structures in the Cydonia region.

The protocol used for the session is described on page 2 of the 1984 CIA document:

The sealed envelope was given to the subject immediately prior to the interview. The envelope was not opened until after the interview. In the envelope was a 3 x 5 card with the following information:

The planet Mars.

> Time of interest approximately I million years B.C.
>
> Selected geographic coordinates, provided by the parties requesting the information, were verbally given to the subject during the interview. [446]

The rest of the CIA document (pp. 3-9) is a transcript of the remote viewer responding to questions about different locations and time periods given to him (In 1984, all known remote reviewers were men). The remote viewer is referred to as "SUB", while the questioner is "MON". This is what the remote viewer reported after being given his first question:

> MON: (Plus 10 minutes, ready to start.)* All right now, using the information in the envelope I've provided, exclusively focusing your attention now, using the information in the envelope, focus on:
>
> 40.89 degrees north
> 9.55 degrees west
>
> SUB: ... I want to say it looks like ah ... I don't know, it sort of looks ... I kind of got an oblique view of a ah ... pyramid or pyramid form. It's very high, it's kind of sitting in a ... large depressed area. [447]

The coordinates were for the Cydonia region, and immediately the remote viewer described some kind of pyramid sitting in a valley. This is remarkable corroboration for multiple researchers who have identified pyramids in the Viking images of this exact region of Mars.

Next, the remote viewer responded to a series of questions concerning the population living in the region shortly before planet-wide geological disturbances occurred, up to a

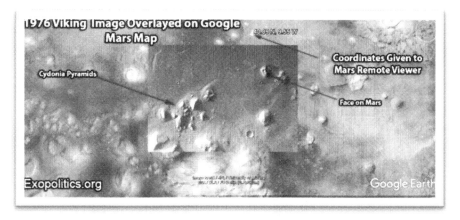

Figure 63. Close up of Cydonia region, Mars

million years ago. He began describing the population's attempt to create a refuge from the disturbances, which included very violent storm activity, by using some kind of stasis or cryogenic technology:

> MON: All right. Go inside one of these and find some activity to tell me about …
>
> SUB: Different chambers … but they're almost stripped of any kind of … furnishings or anything, it's like ah … strictly functional place for sleeping or that's not a good word, hibernations, some form, I can't, I get real raw inputs, storms, savage storm, and sleeping through storms.[448]

What the remote viewer has identified are the pyramid structures on Mars which were used as a refuge where the population planned to sleep through the planetary disturbance in a stasis chamber(s) of some kind.

The remote viewer next described the population as very tall and thin, and gave details about their style of dress:

MON: Tell me about the ones who sleep through the storms.

SUB: ... very ... tall again, very large ... people, but they're thin, they look thin because of their height and they dress like in, oh hell, it's like a real light silk, but it's not flowing type of clothing, it's like cut to fit.[449]

The height and physiology of the Martians suggests that the comparatively low gravity on Mars facilitated gigantism among the human population. The remote viewer next related that the Martian civilization was dying and the population knew this fact:

MON: Move close to one of them and ask them to tell you about themselves.

SUB: They're ancient people. They're ah ... they're dying, it's past their time or age.

MON: Tell me about this.

SUB: They're very philosophic about it. They're looking for ah ... a way to survive and they just can't.[450]

The Martians were waiting to travel somewhere else, yet unknown, to survive. However, a faction was able to escape, presumably off-planet, as the following quote suggests:

MON: What is it they're waiting for?

SUB: They're ... evidently was a ... group or a party of them that went to find ... new place to live. It's like I'm getting all kinds of overwhelming input of the ... corruption of their environment. It's failing very rapidly and this group went somewhere, like a long way to find another place to live.[451]

Next, the remote viewer described what clearly seems to be a spacecraft taking survivors to another planet:

> MON: Okay, when the others left, these people are waiting, when the others left, how did they go?
>
> SUB: ... Get an impression of ... Don't know what the hell it is. It looks like the inside of a larger boat. Very rounded walls and shiny metal.
>
> MON: Go along with them on their journey and find out where it is they go.
>
> SUB: ... Impression of a really crazy place with volcanoes and gas pockets and strange plants, very volatile place, it's very much like going from the frying pan into the fire. Difference is there seems to be a lot of vegetation where the other placed did not have it. And different kind of storm. [452]

The above description is very suggestive of what Earth may have looked like back at that time. So did escaping Martians travel to Earth up to a million years ago leaving behind pyramids and other remnants of their civilization in the Cydonia region, which in modern times have been observed by the Viking Orbiter in 1976 and the CIA's remote viewer in 1984?

In 2004, Joseph McMoneagle, a retired US Army Chief Warrant Officer gave a lecture describing his participation in remote viewing experiments conducted by US Army Intelligence and the Stanford Research Institute.[453] He described the above

Mars incident and even provided additional details confirming that he was the remote viewer cited in the 1984 CIA document. In his lecture, McMoneagle said that the Martians had indeed fled to Earth, and estimated them to be twice the height of a normal human, about 10-12 feet.

In addition, Holmes "Skip" Atwater, who was McMoneagle's training officer at the time, has also discussed the 1984 remote viewing of Mars.[454] In a video presentation, Atwater clarified that the Mars coordinates and instructions were supplied by Hal Puthoff, head of the Stanford Research Institute's remote viewing program, and that the remote viewing session was done under contract to one of the intelligence agencies.[455] With regard to the time period designated in the session, Atwater said that even though the instructions specified remote viewing a million years into the past, McMoneagle was asked to move through the timeline during the session, which meant that the actual catastrophe and exodus on Mars may have occurred tens or hundreds of thousands of years ago, rather than a million. It's important to remember that the fleeing Martians almost certainly carried the stasis chamber technology that their brethren had used on the planet's surface in an attempt to survive the catastrophe.

Another highly significant detail in the CIA document is McMoneagle's reference to a very volcanic environment, where the Martians escaped to using their metallic "boats", or "space arks". As already pointed out, scientists have recently discovered that Antarctica has the highest concentration of volcanoes anywhere on Earth.[456] Volcanic activity today is suppressed by the weight of the kilometers-thick ice sheets, but this was not always the geological situation due to the Earth's shifting poles, as explained by Charles Hapgood in *Path of the Pole*.[457] In the book's Foreword, Albert Einstein expressed his early impression of Hapgood's theory:

> I frequently receive communications from people who wish to consult me concerning their unpublished ideas. It goes without saying that these ides are very seldom possessed of scientific validity. The very first communication, however, that I received from Mr. Hapgood electrified me. His idea is original, of great simplicity, and – if it continues to prove itself – of great importance to everything that is related to the history of the earth's surface.[458]

Einstein went on in the Foreword to explain Hapgood's theory in relation to how the massive ice shelves in the polar region play a critical role in the shifting of the Earth's crust.[459] Hapgood's theory makes it clear that prior to Antarctica being located at the South Pole, it was positioned in a more temperate climatic region, whereby its volcanoes were not suppressed because of vast ice sheets covering them. Consequently, Antarctica's volcanoes were very active during various periods across Earth's geological history. There is good reason, therefore, to speculate that the location which the Martians chose to escape to when fleeing their dying planet was Earth, as McMoneagle himself concluded, and Antarctica in particular as Goode states.

What lends further credence to the arrival of ancient extraterrestrial refugees in spacecraft to a once temperate Antarctica is the testimony of Dr. Pete Peterson. He said he explored the remains of a crashed spacecraft found in Antarctica, and offered details about what he saw in a 2017 interview:

> Well, there was a crash there about 200,000 years ago. And at that time, it was tropical. There were palm trees there. Where the crash was, there were palm trees. That's about three miles under the ice now. That's being excavated. They're putting a huge tunnel down into that one....

So the early stuff, the low stuff, that's down underneath two or three other layers of civilization, is very ancient, seemingly very ancient civilization, probably where they would have used these gear-driven navigation instruments. But the navigation instruments, the gears, the gear ratios, were all set up for this galaxy....

You're going to find several layers, separate layers, of occupation. You're going to find several separate layers of what the exterior was like at the time that civilization was there. Only the last layer was it really Antarctica. Prior to that, it was a tropical island. And part of that, it was part of a very much larger island. Probably good pieces of it were pieces of Atlantis that actually picked up and moved down there.[460]

Peterson's testimony answers the question of where the Martian refugees, described in McMoneagle's remote viewing session, relocated after their planetary catastrophe. As to when the incident occurred, Skip Atwater placed the range from tens of thousands to hundreds of thousands of years ago. Peterson's estimate of 200,000 years is also consistent with McMoneagle's description. Therefore, they all appear to be identifying the same event in Antarctica's history when the continent was located in a tropical zone of the planet. Peterson indicates that there were several landings or crashes of extraterrestrial vehicles, with each corresponding to a different civilization being established, including that of Atlantis.

All of these similarities support Goode's claim that the Pre-Adamite civilization he witnessed, buried deep under the Antarctic ice sheets, indeed belonged to ancient Martian refugees. However, while Peterson and the CIA remote viewing document suggest that the Martians arrived approximately

200,000 years ago, Goode contends they arrived approximately 55,000 years ago via the Moon.

Earth's Moon becomes a Refugee Ship

According to Goode's information, billions of refugees from both Mars and the Super Earth (aka Maldek) found refuge on Earth's Moon, which was another satellite of the Super Earth, before it was destroyed and became the asteroid belt.[461] Goode said the Moon was artificially altered to create vast living areas inside its interior, built out of an ancient cavern system. Support for Goode's extraordinary claim come from a 2017 Japanese scientific study of the moon, which has identified a natural cavern system within it. Here is how Laura Geggel, Senior Writer for *Live Science,* described the findings of the Japanese scientists:

> A city-size lava tube has been discovered on the moon, and researchers say it could serve as a shelter for lunar astronauts. This lava tube could protect lunar-living astronauts from hazardous conditions on the moon's surface, the researchers said. Such a tube could even harbor a lunar colony, they added…. Earth also has lava tubes, but they're not nearly as large as the one discovered on the moon. If the scientists' gravity analyses are correct, the lava tube near Marius Hills could easily house a large U.S. city such as Philadelphia, they said.[462]

The size of the Moon's caverns are large enough to fit a large metropolitan city, as illustrated in the following diagram (Figure 65) showing how Philadelphia could easily fit inside one of them.

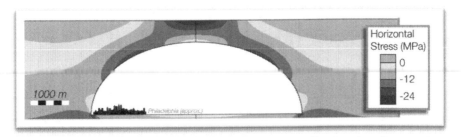

Figure 64. The city of Philadelphia could easily fit inside a theoretical lunar lava tube.
Credit: David Blair/Purdue University

Goode read in the SSP data archive that the Pre-Adamites inhabited the Moon for approximately 440,000 years, and at some point during this period the Moon was moved into its present orbit around the Earth. Compelling verification for this view that the Moon only came to orbit Earth about 60 thousand years ago comes from a number of historical texts. The researcher, Immanuel Velikovsky, found numerous ancient references to pre-lunar cultures on Earth, some of which he described as follows:

> The period when the Earth was Moonless is probably the most remote recollection of mankind. Democritus and Anaxagoras taught that there was a time when the Earth was without the Moon. Aristotle wrote that Arcadia in Greece, before being inhabited by the Hellenes, had a population of Pelasgians, and that these aborigines occupied the land already before there was a moon in the sky above the Earth; for this reason they were called Proselenes.
>
> Apollonius of Rhodes mentioned the time "when not all the orbs were yet in the heavens, before the Danai and Deukalion races came into existence, and only the Arcadians lived, of whom it is said that

they dwelt on mountains and fed on acorns, before there was a moon." Plutarch wrote in The Roman Questions: "There were Arcadians of Evander's following, the so-called pre-Lunar people."[463]

These historical accounts, along with the recent scientific discovery of huge lunar caverns, make Goode's remarkable claim about the Moon being a refugee ship from a destroyed Super Earth very plausible.

Pre-Adamites Escape the Moon and Arrive in Antarctica

The Pre-Adamites (Martians) found refuge on the Moon, but eventually another conflict came and they had to quickly move to the nearby Earth. According to Goode, they chose Antarctica to rebuild their civilization, due in part to the presence here of "Ancient Builder Race" technologies which were still functioning:

> Then they [Pre-Adamites] end up on the Moon for a period of time. And then somehow they ended up getting chased off the Moon. There were some attacks that occurred, and after that they fled, but their craft were too damaged to leave our solar system or make it to another planetary sphere. So since they had to crash-land on Earth, they decided that they would go to this one continent that still had working Ancient Builder Race technology that was ... 1.8 billion years old.[464]

After a crash landing in Antarctica, the Pre-Adamites only had their three motherships to use to establish a new colony on Earth.

It was the technology in these motherships that gave the Pre-Adamites hope that they could rebuild their civilization:

> Well, they only had the technology that they had on these three craft with them. That's all the technology they had. So they had to cannibalize and repurpose that technology from the spacecraft once they had crash-landed.[465]

David Wilcock interviewed Goode on *Cosmic Disclosure* about these ancient events, and had some corroborating information to share from Dr. Pete Peterson regarding the size and location of one of the discovered motherships:

> David: And I want to point out, that this was one of the absolutely stunning details in which I start to ask Pete [Peterson] on the phone, "Do you know anything about Antarctica?"
>
> And he independently says, "They're going to announce that they found a mothership." He only knew about one – a mothership that was 30 miles wide, mostly circular in shape – that's been found under the ice. [466]

In contrast to Peterson, Goode says that the largest of the motherships was three miles in length.

During that interview, Goode described the Pre-Adamites who settled in the Antarctic: "They range 12-14 foot tall. They have elongated skulls. They were very spindly, thin."[467] Significantly, this description is very similar to how the inhabitants on Mars were described in the May 1984 CIA remote viewing document.[468] After spending over four hundred thousand years on the Moon with its even lower gravity field, the Pre-Adamites may have increased in body height even further to finally result in their spindly 12-14 foot stature.

On Earth, because of stronger gravity, the Pre-Adamites would now find themselves at a clear disadvantage compared to native inhabitants, in terms of physical strength, speed and stamina. In turn, in terms of geopolitical power, this meant the Pre-Adamites would have to rely on their advanced technologies, along with those more advanced 'Builder Race' technologies located in Antarctica. Consequently, references to Atlantis and other technologically advanced cultures found around the planet may be signs of a dominant global civilization established by the Pre-Adamites, whose hub was Antarctica. Indeed, the Pre-Adamites may well have been the fabled "Fallen Angels" described in the Book of Enoch, whose advanced technologies led to a major conflict in the pre-diluvial world.

CHAPTER 14

Antarctica, Demigods &
the Book of Enoch

Extraterrestrial Colony Created Elite Bloodline Rulers

After arriving in Antarctica from the Moon approximately 55,000 years ago, the Pre-Adamites found a rich fertile planet with a large native population less technologically advanced than themselves. Soon after, the Pre-Adamites decided to create hybrid beings using their advanced genetic technology, who would act as intermediaries for them with the rest of the Earth's population, as Corey Goode explained:

> They had created hybrids because they could not operate in our environment very well. And they created hybrids of them and the humans that were here on Earth ... all of the main Pre-Adamites that were pure blood were down in Antarctica ... There was a group of these Pre-Adamites, of this Pre-Adamite bloodline, that was in the Central America, South America region, and there was another completely different bloodline group – both royals – in Asia and Europe.[469]

Essentially, while the pure-blooded Pre-Adamites remained in Antarctica close to the advanced technologies in their large motherships, the hybrid Pre-Adamites were to be used to rule over humanity in the different colonies established around the planet.

Goode's historical information parallels that of the Ancient Egyptian historian, Manetho, who wrote about pre-dynastic times – when the "Gods" directly ruled over Egypt. These Gods then placed their genetic offspring, hybrids or demi-gods into leadership positions. Manetho's writings are preserved in various historical fragments retold by ancient writers such as Eusebius, who wrote:

> From the Egyptian History of Manetho, who composed his account in three books. These deal with the Gods, the Demigods, the Spirits of the Dead, and the mortal kings who ruled Egypt down to Darius, king of the Persians ... After the Gods, Demigods reigned for 1255 years, and again another line of kings held sway for 1817 years: then came thirty more kings of Memphis, reigning for 1790 years; and then again ten kings of This, reigning for 350 years.... [470]

It should be noted that ancient depictions of Egyptian rulers such as the Emperor Akhenaton and Queen Nefertiti show them possessing elongated skulls. Goode described the Pre-Adamites in Antarctica as possessing elongated skulls, which held larger sized brains. Scientists today point out brain size capacity is a direct marker of intelligence, and consequently, of leadership abilities.

In 1913, elongated skulls were discovered in the town of Boskop, South Africa which showed that a race of beings once inhabited that area who had a 30% larger brain capacity than modern Homo sapiens. The Boskop remains, and similar discoveries of elongated skulls at sites around the world have led scientists to postulate how much smarter this group would have been compared to normal Homo sapiens. Two distinguished neurosurgeons published a book about the Boskop skull discovery, which casts light on the relationship between brain size and intelligence.[471] In an extract published in *Discover Magazine,* they wrote:

> Even if brain size accounts for just 10 to 20 percent of an IQ test score, it is possible to conjecture what kind of average scores would be made by a group of people with 30 percent larger brains. We can readily calculate that a population with a mean brain size of 1,750 cc would be expected to have an average IQ of 149.

> This is a score that would be labeled at the genius level. And if there was normal variability among Boskops, as among the rest of us, then perhaps 15 to 20 percent of them would be expected to score over 180. In a classroom with 35 big-headed, baby-faced Boskop kids, you would likely encounter five or six with IQ scores at the upper range of what has ever been recorded in human history.[472]

It is fair to conclude, based on scientific studies, that humans possessing elongated skulls with larger brain capacity were the natural leaders of the ancient world. As members of the ruling human Egyptian dynasties, Akhenaton and Nefertiti therefore possessed a genetic trait that could be traced to the original gods, the Pre-Adamites, who were distinguished by their notable height and cranial elongation. This genetic or "bloodline" connection to the Gods/Pre-Adamites, as claimed by Goode, is where the ancient idea of the divine right of kings was derived.

Figure 65. A detail from a wall-painting picturing the daughters of Pharaoh Akhenaten and Queen Nefertiti, 18th Dynasty, c.13-45-1335 BC

A similar scenario is described in the Sumerian King's List, which chronicles the reigns of kings both prior to, and after, the Great Flood, an event also depicted in the Old Testament and in Sumerian texts, such as the *Epic of Gilgamesh*.[473] The account of the first list of kings begins:

> After the kingship descended from heaven, the Kingship was in Eridug. In Eridug, Alulim became king; he ruled for 28800 years. Alaljar ruled for 36000 years. 2 kings; they ruled for 64800 years. Then Eridug fell and the kingship was taken to Bad-tibira. In Bad-tibira, En-men-lu-ana ruled for 43200 years. En-men-gal-ana ruled for 28800 years. Dumuzid, the shepherd, ruled for 36000 years. 3 kings; they ruled for 108000 years. ..."[474]

Here the King's List is clearly suggesting that the kings who came before the Great Flood were connected to heavenly beings, and their fantastically long periods of rule also indicate these kings were considered as Gods or Angels, whom we would likely call today, extraterrestrial visitors.

In the case of the postdiluvian rulers, the King's List continues:

> "After the flood had swept over, and the kingship had descended from heaven, the kingship was in Kic. In Kic, Jucur became king; he ruled for 1200 years. Kullassina-bel ruled for 960 (*ms. P2+L2 has instead:* 900) years. Nanjiclicma ruled for (*ms. P2+L2 has:*) 670 (?) years. En-tarah-ana ruled for (*ms. P2+L2 has:*) 420 years"[475]

This passage lays the foundation once again for the divine right of kings resting in some connection to heavenly beings. This time, however, the rulers are described as reigning for much shorter periods than the original kings, yet much longer than human kings. This passage suggests these kings were demigods, as described by Manetho, or human-extraterrestrial hybrids as described by Goode, who could live up to a thousand years or so.

As the Kings List progresses, the reigns of rulers becomes shorter and shorter, as typified by Sargon of Agade who ruled between 2340-2284 BC:

> In Agade, Sargon, whose father was a gardener, the cupbearer of Ur-Zababa, became king, the king of Agade, who built Agade (*ms. L1+N1 has instead:* under whom Agade was built); he ruled for 56 (*ms. L1+N1 has instead:* 55) (*ms. TL has instead:* 54) years. Rimuc, the son of Sargon, ruled for 9 (*ms. IB has instead:* 7) (*ms. L1+N1 has instead:* 15) years. Man-icticcu, the older brother of Rimuc, the son of

Sargon, ruled for 15 (*ms. L1+N1 has instead:* 7) years. Naram-Suen, the son of Man-icticcu, ruled for (*mss. L1+N1, P3+BT14 have:*) 56 years. Car-kali-carri, the son of Naram-Suen, ruled for (*ms. L1+N1, Su+Su4 have:*) 25 (*ms. P3+BT14 has instead:* 24) years. [476]

By Sargon's time, human kings are clearly being identified with similar life spans as modern humans. But what occurred that led to this remarkable transition, from God-kings (extraterrestrials) ruling thousands of years, demigods (human-extraterrestrial hybrids) ruling centuries, and then human kings (with genetic connections to their predecessors) ruling mere decades?

Both Manetho's History and the Kings List indicate that there was a steady decrease in the genetic purity, and hence, longevity of the rulers of ancient civilizations, dating back hundreds of thousands of years. Certainly, the Christian Old Testament reports a similar phenomenon; insofar as pre-diluvial figures such as Adam, Jared, Methuselah and Noah who all lived to be over 900 years of age. [477]

In the immediate period after the great flood, the Old Testament refers to patriarchal figures living steadily shorter lives. For example, Shem, one of the sons of Noah lived to 600 years, his son Arpachshad lived to be 438 years of age, and his grandson Shelah lived to be 403 years old. [478] By the time of Amram, father of Moses (c. 1400 BC), lives were much shorter: Amram lived to 137 and Moses lived to 120.

Conventional scholars dismiss these extensive periods of time chronicled in ancient sources as myths. Goode's information suggests otherwise. One or more groups of extraterrestrial visitors settled on Earth in various historical epochs, first ruling over, and then interbreeding with, humanity to create the successive dynasties of kings, emperors and biblical patriarchs. It's quite probable that the first kings of Egypt, Sumer, etc., either derived their authority and power from the Martians/Pre-Adamites who had settled in Antarctica, or were Martians/Pre-Adamites themselves.

The Great Flood that occurred approximately 10,000 B.C.E. appears to have been a very significant event in the transition from the age of God-kings, to the successive ages of Demigod kings, and finally, human kings and patriarchs. As the Kings List makes clear, the Great Flood is the historical marker for the transition from an age of direct rule by the Gods (who individually ruled thousands of years), to Demigods (who ruled for centuries). This is mirrored very closely in the Old Testament in terms of ancient patriarchs both prior to and after the Great Flood.

Apparently, the Great Flood depicted in the Sumerian, Biblical and other world traditions was a catalyst for this very significant transition in human history. Graham Hancock, in his book *Magicians of the Gods,* describes various possible explanations for the last great flood, such as successive asteroid impacts.[479] The scientific data Hancock provides is very impressive in detailing how the Great Flood depicted in historical texts may have been caused by multiple asteroid impacts spanning the Younger Dryas period from c. 12,900 to 11,700 calendar years ago.

Charles Hapgood's theory of a pole shift, however, offers a more compelling explanation for the last Great Flood. Not only does a Pole Shift explain how the last Great Flood may have occurred, but it also provides a very important way of understanding the transition in ages from the direct rule by Gods, to that of Demigods and humans, as depicted in the Sumerian Kings List and Manetho's history of Egypt.

If the Pre-Adamites did crash-land in Antarctica 55,000 years ago and establish their technological hub there as Goode claims, then we have an understanding of why the last Great Flood/Pole Shift marked a decisive turning point in human civilization. Antarctica, being repositioned to the polar region, now lay under ice, which in years quickly thickened to grow into massive ice sheets. The Pre-Adamites/Martians could do little more than ride this cataclysmic event out using the stasis chamber technology brought over from Mars aboard their motherships, as illustrated in McMoneagle's May 1984 Mars remote viewing session discussed in chapter 13. It's also likely that these stasis chambers enabled the Pre-Adamites/Martians to live the tens of thousands of years described in the Kings List.

Not only did the Great Flood bury the mother colony of the Pre-Adamites civilization under ice, but it cut off their hybrid vassals from the advanced technologies used to maintain control over their colonies around the Earth. This led to the Pre-Adamites' major extraterrestrial rival, who vied for direct influence over human affairs, to now step forward to fill the vacuum.

The Pre-Adamites were not the only extraterrestrial race active on Earth 55,000 years ago according to Goode's sources. He says that among the races was a non-human species, the Reptilians, who quickly emerged as the Pre-Adamites' primary rival in dominating planetary affairs. Goode details the conflict between the Pre-Adamites and the Reptilians; and how the Pre-Adamites in the beginning had the upper hand, due to their advanced technologies. However, the Reptilians slowly gained ascendance after a series of "smaller catastrophes" took place as a precursor to the even more destructive geological event, the Great Flood, which came later:

> Apparently these Pre-Adamites have been in conflicts with the Reptilians for a while. These Pre-Adamites they stated were not good guys at all, but were in some sort of a conflict with the Reptilians, and had actually kept the Reptilians in check here on the Earth during that time that they had crash-landed here.
>
> There had been a couple other smaller catastrophes that happened where they had lost their power and the Reptilians always find an

opportunity to come back in a moment of weakness.[480]

During the great catastrophe (approximately 12,000 years ago), the Antarctic continent was flash frozen during the Earth's sudden pole shift, which cut off the Pre-Adamites' outposts around the world from their main base of power.

Figure 66. Graphic depiction of Corey Goode visiting Antarctica excavations with Pre-Adamite/hybrid bodies. Courtesy of www.Gaia.com

Now, the Reptilians would become dominant in planetary affairs:

> ... But after these cataclysms that occurred on Earth, the Pre-Adamites and the Reptilians sort of had a truce or a treaty. And after that point, the Reptilians pretty much controlled all of Antarctica and the Pre-Adamites had zero ability to get access to their ancient technology, their libraries. Everything was down there.

Hybrids of the Pre-Adamites were able to escape the catastrophe in their global outposts, but those of the pure bloodline were stuck in Antarctica in their giant motherships.

This [Pre-Adamite] civilization controlled the entire planet. What little resources they had, they were able to control the planet. After this last cataclysm occurred, none of the survivors, Pre-Adamite survivors, had access to their technology.

So we mentioned the group that was in Asia, Pre-Adamite group, and there was another one in South America, Central America, they could no longer visit or communicate with each other. They were separated.[481]

The elongated skulls found in South and Central America trace back to the Pre-Adamite hybrids who ruled over their colonies in these areas:

In South and Central America is where they had set up most of their enclaves. And they had been set up around other Pre-Adamite structures that were now destroyed because of the cataclysm. There were huge earthquakes that basically liquified the ground and a lot of the buildings, massive buildings, they had just fell apart and fell and sunk into the ground. They were running the hemisphere. They were mixing their genetics with some of the indigenous people of South and Central America. That's why we have elongated skulls beings that have a different colored skin but different genetic mix.[482]

The two major Pre-Adamite colonies, one spanning Asia and Europe, and the other encompassing South and Central America, vied with each other in a competition between their two bloodlines, which continues today through two major factions of the Illuminati:

> And they had always had some sort of a competition between these two bloodline groups even before the cataclysm. This whole bloodline of these Cabal or Illuminati-type people, they trace their bloodlines through these Pre-Adamites. [483]

The rivalry between the Pre-Adamite bloodline factions and their conflict with the Reptilians provides an answer to the question of what happened to the hybrids with elongated skulls that formed humanity's ruling elite.

In their book, *Big Brain,* Gary Lynch and Richard Granger attempt to answer the question of what happened to the Boskop humans, whose elongated skulls held larger brains, giving them intelligence significantly higher than modern humans. At some point, they formed the leadership class, as Egyptian artwork clearly depicts, yet by the modern era they had disappeared. Lynch and Granger ask the question, why did this childlike large-skulled human disappear?:

> The Boskops coexisted with our Homo sapiens forebears. Just as we see the ancient Homo erectus as a savage primitive, Boskop may have viewed us in somewhat the same way. They died and we lived, and we can't answer the question why. Why didn't they outthink the smaller-brained hominids like ourselves and spread across the planet?[484]

Given what Goode has revealed about the Reptilian/Pre-Adamite conflict, we can speculate that post-diluvian Homo sapiens were inspired by the Reptilians to revolt against the Pre-Adamite hybrids. This is echoed in the Old Testament, in the story in which Adam and Eve are tempted by the serpent to eat of the tree of knowledge, and are expelled out of Eden. After the Great Flood, surviving Homo sapiens subsequently conducted ancient pogroms to kill most of the Pre-Adamites hybrids (Boskops) and drive the survivors underground.

If we accept that Reptilians are another extraterrestrial group who exercised great influence from behind the scenes through their own hybrids, then we can see that the Earth's hidden rulers are divided into distinct factions, with a long history of conflict through the use of human proxies. This is perhaps nowhere better exemplified than in the "Book of Enoch", which describes different groups of "angels" at war with one another. In a *Cosmic Disclosure* interview, Goode and Wilcock exchanged the following on this subject:

> David: So in the Book of Enoch, they're describing this group as the fallen angels.
>
> Corey: Uh-huh.
>
> David: So you're saying these Pre-Adamites with the elongated skulls, that that is the fallen angel storyline.
>
> Corey: It is the fallen angel storyline, yes. Many of the original refugee Pre-Adamites are currently in stasis in their motherships buried under the Antarctica ice ... Well, they had a number of beings that were in stasis. The information I received was that the surviving Pre-Adamites, the bloodline that originally came from another planet, had put themselves in stasis before this cataclysm occurred about 12,800 years ago. They have not awoken

them yet. They're trying to decide what they're going to do.[485]

What Goode has been told and personally witnessed in Antarctica tells us a lot about Antarctica's secret history, and the role of different extraterrestrial groups who established control over this vast icy continent almost double the size of the lower 48 U.S. states. Goode also leaves us with a profound question to contemplate: "What happens when the Pre-Adamites are awakened in their stasis chambers and discover that our current global civilization is very different to what they may desire?"

Antarctica & Imprisoned Fallen Angels from the Book of Enoch

On March 14, 2017, Israeli News Live published a provocative story titled, "The Fallen Angels Imprisoned in Antarctica and are still Alive."[486] The commentator, Steven Ben-Nun, a Political Journalist and Biblical Scholar, analyzed the apocryphal *Book of Enoch*, which describes the experiences of Enoch, a pre-deluvian biblical figure who was taken into the heavens to witness and play a key role in a major celestial conflict. Since Goode asserts that the Fallen Angels information in the Book of Enoch is relevant to the Pre-Adamite discovery in Antarctica, it's worth exploring this subject further.

According to the ancient text, Enoch became the principal intermediary between two sides in a conflict between "Fallen Angels", and "Righteous Angels" serving an all-knowing deity, referred to as "the Lord". Multiple issues incited the conflict, including the Fallen Angels' practice of interbreeding and/or performing genetic experiments upon humanity, and their passing on of forbidden knowledge and technologies to the still developing human civilization.

The *Book of Enoch* begins with the arrival of 200 Fallen Angels in the area of Mount Hermon, which borders modern-day Lebanon and Syria. The Fallen Angels began interbreeding and/or genetically modifying the local inhabitants.

> 6.1 And it came to pass, when the sons of men had increased, that in those days there were born to them fair and beautiful daughters.
>
> 6.2 And the Angels, the sons of Heaven, saw them and desired them. And they said to one another: "Come, let us choose for ourselves wives, from the children of men, and let us beget, for ourselves, children." …
>
> 6.6 And they were, in all, two hundred and they came down on Ardis, which is the summit of Mount Hermon. And they called the mountain Hermon because on it they swore and bound one another with curses.[487]

While the Fallen Angels had established an outpost on Mt. Hermon, it was Antarctica where they would ultimately be confined after losing the heavenly battle with the Righteous Angels, according to Ben-Nun's analysis of the *Book of Enoch*.

Ben-Nun cites passages from the *Book of Enoch* which are very suggestive of Antarctica being the very location where Enoch was taken to witness celestial events:

> 18.5 And I saw the winds on the Earth which support the clouds and I saw the paths of the Angels. I saw at the end of the Earth; the firmament of Heaven above.
>
> 18.6 And I went towards the south, and it was burning day and night, where there were seven

mountains of precious stones, three towards the east and three towards the south.

18.7 And those towards the east were of coloured stone, and one was of pearl, and one of healing stone; and those towards the south, of red stone.

18.8 And the middle one reached to Heaven, like the throne of the Lord, of stibium, and the top of the throne was of sapphire.[488]

What's interesting in the above passage is that Enoch refers to a location that "was burning day and night". Ben-Nun believes this fits the conditions of Antarctica during the Southern Hemisphere summer season when there is 24 hour sunlight.

Regarding the seven mountains mentioned in the book's passages, the one that "reached to Heaven" appears to refer to Mount Vinson in the Sentinel Range of Antarctica, according to Ben-Nun. He also refers to six nearby mountains in that range which might qualify as the other mountains described in the ancient text. Mount Vinson is the highest mountain in Antarctica, and is located towards the middle of the Sentinel Range. It would have stood out with its massive height of over 16,000 ft. (4892 m) just as spectacularly in ancient times as it does today.

Ben-Nun speculates the southern and eastern alignment of the six adjacent mountains to the central mountain, described in the *Book of Enoch*, compares to Mount Vernon and it's six highest adjacent mountains' alignment *prior* to the catastrophic flood event (which coincided with a shifting of the Earth's axis of rotation).

This in turn directly relates to the research conducted by Sir Charles Hapgood, whose proposed pole shift happened at the end of the last ice age, about 11,000 BC.[489] Ben-Nun's conjecture is interesting, but as one can see from the map showing the Sentinel Range where Mount Vinson is situated (see Figure 67),

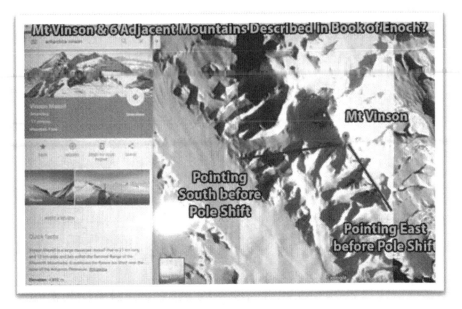

Figure 67. Mount Vinson and Adjoining Mountains

there are far more than seven mountains in the range. Thus, Ben-Nun's theory is not conclusive, as he himself points out. Nevertheless, he gives us the possible location of the imprisoned Fallen Angels; Mount Vinson, and/or six other mountains in the Sentinel Range.

Ben-Nun goes on to discuss the *Book of Enoch's* reference to imprisoned Fallen Angels, and concludes they were removed from Mount Hermon and relocated to Antarctica (Mount Vinson/Sentinel Range):

> 8.14 And like a spirit questioning me, the Angel said: "This is the place of the end of Heaven and Earth; this is the prison for the Stars of Heaven and the Host of Heaven.

> 18.15 And the stars which roll over the fire, these are the ones which transgressed the command of the Lord, from the beginning of their rising,

because they did not come out at their proper times.

18.16 And He was angry with them, and bound them until the time of the consummation of their sin, in the Year of Mystery."[490]

Ben-Nun's reference to the Fallen Angels still being alive in their Antarctica prison raises an intriguing possibility. Is the foretold liberation of the Fallen Angels in the "Year of Mystery" an event we will shortly witness?

For an answer, some intriguing parallels between Ben-Nun's study of the *Book of Enoch*, and the recent disclosures by Corey Goode concerning Antarctica need to be analyzed. Like the pre-diluvial biblical figure Enoch, Goode also says he was taken to witness celestial events by a recently arrived group of highly evolved extraterrestrials – called the "Sphere Being Alliance". Again, like Enoch, Goode was purportedly asked to act as an intermediary between two sides in a "heavenly" or solar system-wide conflict.

Goode claims this all began in March 2015, when he accepted the role of delegate for the Sphere Being Alliance in on-going negotiations between different factions and groups from both Earth and off-world civilizations.[491] Events developed over the next two years that led to Goode being taken to Antarctica twice to witness events. Goode says the most recent in January 2017 involved a clandestine trip enabling him to view the remains of a Pre-Adamite civilization that had flourished until a major global catastrophe, corresponding with the great flood, had destroyed coastal cities and low-lying lands all around the world approximately 12,000-13,000 years ago.[492]

Goode explains that the Pre-Adamites survived by entering stasis chambers in the largest of three motherships that had crash landed on Earth around 60,000 years ago, after the Pre-Adamites were expelled from the Moon due to a conflict with the dominant extraterrestrial factions there. The Pre-Adamites set up their major base of operations in Antarctica, and established outposts in Asia, Europe and the Americas.[493] Goode further contends that a conflict soon emerged between the Pre-Adamites and other human-looking extraterrestrial groups, who had been conducting 22 specific genetic experiments involving surface humanity for nearly 500,000 years.

Goode first discussed the 22 long-term genetic experiments at length in a May 17, 2015 *Facebook* post:

> In the last Human-Like Federation Conference ... after a short presentation I was asked to deliver there was then a follow up of each of the over 60 Beings Present (Usually Never More Than 40 Beings Present). Among these were those involved in the 22 Scientifically Documented Presentations delivered decades prior claiming to have created/engineered Earth Humans genetically. Each of the beings present gave a short presentation of what they considered their "Contributions To Humanity" over the thousands of years they had been managing human affairs. This was the exact same setting that I had attended in an Intuitive Empath Support Role when I was younger and did not understand what was going on. Being an actual "Delegate" this time I understood everything going on and said. It was interesting that each group seemed to make a case that they had been a positive influence on the creation and management of humanity. Of course this was their point of view and the "IE's" present in support of us detected no danger or deception.[494]

Alex Collier, who claims he had interactions with a group of extraterrestrials from the Andromeda constellation, says they told him about 22 genetic experiments run by different human-looking alien groups stretching back into human antiquity:

> In a nutshell, we are a composite of a lot of different races, 22 to be exact. It is a physiological fact that there are 22 different body types on the planet. And that is the result of the extraterrestrial races.[495]

Goode and Collier's historical accounts suggest that these human-looking extraterrestrial groups eventually came into conflict with the Pre-Adamites operating out of Antarctica.

Here, Goode's explanation of historical events resonates with Ben-Nun's own analysis of events found in the *Book of Enoch*. Goode's Pre-Adamites appear to be the same "Fallen Angels" described in the age-old book in terms of their interbreeding and/or genetic experiments using the local human population. Also, the ancient text's landing of 200 Fallen Angels at Mount Hermon and the account of their subsequent actions is consistent with Goode's claim that the Pre-Adamites established colonies all over the Earth, genetically altered the local humans, and installed hybrids into leadership positions.[496]

Regarding the Righteous Angels described in the *Book of Enoch*, these appear to be the extraterrestrial groups who were conducting the 22 long-term genetic experiments, which were later interfered with by the Pre-Adamites. Recently, the Sphere Being Alliance has leveled the playing field for constructive negotiations between different space programs and extraterrestrial factions, the modern-day Fallen and Righteous Angels, and in these terms, appears to play a similar role to the all-knowing "Lord" mentioned in the *Book of Enoch*. Ben-Nun's analytical conclusion that the Fallen Angels were imprisoned in Antarctica, and are still alive, corresponds with Goode's claim that the Pre-Adamites are in stasis-chambers on one of their three spacecraft buried deep below the Antarctic Ice Shelf. Finally, Goode has revealed that excavations in Antarctica are underway, and that the stasis chambers containing the Pre-Adamites have been found.[497] He says that the decision about whether to awaken them or not is still to be made.[498]

The hybrid descendants of the Pre-Adamites, who are among the elite bloodline families that secretly have ruled humanity, are eager to gain access to the stasis chambers and to reawaken their ancient ancestors. This may explain why there was a succession of VIP visitors to Antarctica in 2016/2017, including former Secretary of State John Kerry, former astronaut Buzz Aldrin, Russian Patriarch Kirill, Australia's Governor General Sir Peter Cosgrove, and others.

The goal of reawakening the Pre-Adamites appears to be to have them re-assume their former dominance in planetary affairs, thereby allowing their hybrid off-spring to step out of the shadows and directly rule over humanity. This suggests that the "Year of Mystery" mentioned in the Book of Enoch, when the "Fallen Angels" are liberated, may be very close.

All this leads to an intriguing question. Is Goode himself a modern-day Enoch being given the key role to witness and mediate between different human and extraterrestrial factions as the delegate of the Sphere Being Alliance, which appears to be functioning identically to the all-knowing deity described in the *Book of Enoch*? If so, this would suggest that Enoch himself was an extraterrestrial contactee of the Sphere Being Alliance or a similar group, 13,000 years ago. If the Pre-Adamites/Fallen Angels are awakened/liberated, then there will certainly be a need to deal with them and their hybrid offspring regarding the roles they are to play as humankind awakens to the truth of its history, including its manipulation by multiple extraterrestrial groups and elite bloodline families.

Conclusion

According to the history shared by Goode and Tompkins, there are three primary factions of extraterrestrials who have interacted with humanity over the millennia and intersected in Antarctica's ancient past. The first are the human-looking or "Nordic" extraterrestrials. They have been responsible for 22 long-term genetic experiments going back 500,000 years, where human genetics have been altered in accord with global contingencies. Apparently, according to Goode, some of these alterations are connected to the use of ancient "Builder Race" technologies, and to Inner Earth beings who claim they are our future descendants sent back in time 18 million years or so to protect the timelines.

Antarctica is a vast repository for some of this ancient builder race technology, which is highly sought after or protected by the Nordics and Inner Earth groups. Both of these groups appear to support human freedom and creativity as necessary ingredients for human evolution, which appears to occur in planet wide experiments that last millennia before being reset through some major global cataclysmic event. This corresponds to historical references by the Ancient Greeks, Mayans, Vedic Indians, Native Americans, etc., about the different planetary ages when great advances in human civilization occurred. This rather benign view of human evolution does not appear to be shared by the other two major extraterrestrial factions, who ruthlessly use humanity as proxies for expanding their power and influence.

The second faction are the Pre-Adamites who crash-landed in Antarctica approximately 55,000 years ago per Goode's testimony, and established colonies around the planet where their hybrids were put in positions of power. The human colonies, controlled by the Pre-Adamites, appear to have been slave societies, wherein a strict hierarchical power system was put in place and a person's position depended on their bloodline. This is the origin of today's elite bloodline families who manipulate humanity from behind the scenes.

The *Book of Enoch* describes the conflict between the Pre-Adamites and the "Nordics", which led to the former's banishment/imprisonment in Antarctica. This suggests that the Pole Shift which flash froze Antarctica was a result of geo-engineering by the Nordics or higher dimensional beings that wanted to free humanity from Pre-Adamite domination, for a new planet-wide experiment.

The third extraterrestrial faction are the Reptilians, whose history on Earth also goes back many millennia. It's quite possible that the end of the Dinosaur age 65 million years ago was another geo-engineered event that drove the surviving Reptilians underground. Once again, this may have been a result of the Nordics and/or higher dimensional beings using geo-engineering in order to reset the surface conditions for a new set of global genetic experiments, this time allowing the evolution of humans on the planet's surface.

The Reptilian presence in Antarctica was made possible by the cataclysmic events that occurred approximately 12,000 years ago, which removed their major global rival from the scene and drove them underground – the Pre-Adamites. With the eclipse of the Pre-Adamites, the Reptilians were able to augment their power, and asserted their influence around the planet through endless wars of conquest using human proxies. This is certainly the scenario depicted in books such as Jim Marrs' *Rule by Secrecy* and William Bramley's *Gods of Eden*.

Earlier in this book, it was shown how a faction of Reptilian extraterrestrials called "Draconians" had established a presence in Antarctica within many large caverns, and later shared the location of several of them with the German secret societies during WWII. This founded a German-Draconian Reptilian alliance that has steadily grown, through secret agreements with U.S. and European leaders, into a global power – the Fourth Reich.

Currently, Antarctica is a power base for the Fourth Reich through the presence of two inter-linked space programs.

The first program, the Dark Fleet, is the direct offshoot of Hitler's and German Secret Societies' flying saucer projects, which later allied with Draconian Reptilians based in Antarctica. The second is the Interplanetary Corporate Conglomerate, which grew out of the secret agreements between the Fourth Reich and the Eisenhower administration. These agreements allowed for the rapid expansion of the German Antarctic bases into vast industrial areas, six of which Goode purportedly saw first-hand during his two trips to Antarctica in 2016 and 2017.

Today, Antarctica is largely controlled by the Fourth Reich/Dark Fleet and their partners, the Interplanetary Corporate Conglomerate — both of whom are aligned with the Draconian Reptilians. This raises the question: What will happen if the Reptilians former enemy, the Pre-Adamites, are awakened from their stasis chambers by their Cabal/Illuminati descendants? Will this lead to outright conflict causing planetary devastation, or even an "alliance of convenience" between the Reptilians and Pre-Adamites to subjugate the rest of humanity? Information from both Goode and Tompkins suggests that the Fourth Reich's practice of slave labor, which is extensively applied in Antarctica, is likely to be embraced by the Pre-Adamites. Such shared ideologies could catalyze an alliance of convenience to enslave the rest of humanity, who would become mere subjects for their own long-term genetic experiments. In the next chapter, I will reveal the extent of human experimentation taking place in Antarctica, and how this needs to be exposed in order to prevent such abhorrent practices from being introduced across the rest of the planet by stealth.

CHAPTER 15

Illegal R&D in Antarctica

Unacknowledged Special Access Programs & the Antarctic Treaty

When it comes to any classified projects in Antarctica that violate the 1961 Antarctic Treaty, there are two types of "unacknowledged" research and development systems that need to be considered. The first involves unacknowledged programs that are related to different national militaries and intelligence agencies, which are allocated at various levels of security classification for the conduct of such programs. Within the U.S. Department of Defense and the U.S. Intelligence Community, such classified programs are known as Special Access Programs (SAP). Some of these classified programs are "unacknowledged" in so far as their existence is not publicly admitted.

According to a 1995 Department of Defense manual titled, "National Industrial Security Program Operating Manual":

> There are two types of SAPs, acknowledged and unacknowledged. An acknowledged SAP is a program which may be openly recognized or known; however, specifics are classified within that

SAP. The existence of an unacknowledged SAP or an unacknowledged portion of an acknowledged program, will not be made known to any person not authorized for this information.[499]

The Department of Defense manual goes on to clarify the measures taken to keep the existence of unacknowledged programs secret:

> Unacknowledged SAPs require a significantly greater degree of protection than acknowledged SAPs ... A SAP with protective controls that ensures the existence of the Program is not acknowledged, affirmed, or made known to any person not authorized for such information. All aspects (e.g., technical, operational, logistical, etc.) are handled in an unacknowledged manner.[500]

Stringent security requirements pertain to an Unacknowledged SAP (USAP), however, it may even be further classified, making it a "Waived" USAP. According to a 1997 Senate investigation:

> Among black programs, further distinction is made for "waived" programs, considered to be so sensitive that they are exempt from standard reporting requirements to the Congress. The chairperson, ranking member, and, on occasion, other members and staff of relevant Congressional committees are notified only orally of the existence of these programs.[501]

A strictly oral briefing is given for this type of program that is so classified that U.S. Congressional members and others learning about it cannot admit to its existence or consult others for expert advice, which means that USAPs have no effective congressional oversight placed upon them. Congress has to accept *the word* of the sponsoring military service or intelligence agency that the program is being run responsibly and in accord with U.S. laws and the Antarctic Treaty. Further, any Congressional member is required to deny the existence of such a program, and instead refer to a "cover story". In this regard, a 1992 supplement to an earlier version of the Department of Defense manual states:

> Program Cover stories. (UNACKNOWLEDGED Program). Cover stories may be established for unacknowledged programs in order to protect the integrity of the program from individuals who do not have a need to know. Cover stories must be believable and cannot reveal any information regarding the true nature of the contract. Cover stories for Special Access Programs must have the approval of PSO [Program Security Officer] prior to dissemination.[502]

The second type of program dealing with unacknowledged activities involves private corporations. Corporate run USAPs invoke similar security procedures to those used by the U.S. military or intelligence agencies as a condition for working on highly classified contracts. These industry standard security procedures by the Department of Defense are outlined in the "National Industrial Security Program Operating Manual". Here is how the 1997 U.S. Senate Report summarizes the current situation:

> Industrial contractors performing classified contracts are governed by the National Industrial Security Program (NISP), created in 1993 by Executive Order 12829 to "serve as a single, integrated, cohesive industrial security program to protect classified information." A Supplement to

the NISP operating manual (NISPOM) was issued in February 1995 with a "menu of options" from which government program managers can select when establishing standards for contractors involved with special access programs.[503]

It is common for national militaries and intelligence agencies to award contracts to private corporations for running aspects of unacknowledged programs, including those related to activities in Antarctica and Outer Space. For example, Lockheed Martin, Raytheon, Science Applications International Corporation (SAIC), and General Dynamics are a few of the many U.S. companies that are awarded military/intelligence contracts for research and development on highly classified programs, which include operations in Antarctica.

Other Antarctic Treaty signatory states have their own equivalents to USAPs, involving unacknowledged activities conducted by national militaries and intelligence organizations, along with the assistance of private corporate contractors. The scope and budgets of these officially sanctioned unacknowledged programs are not well known, and traditionally, are part of the nebulous world of "black programs" that are funded in opaque ways, as explained by Tim Cook in *Blank Check: The Pentagon's Black Budget*.[504]

The Pentagon's "deep black budget" was estimated to be as high as $1.7 trillion per year by the end of the Clinton administration in January 2001.[505] To fully appreciate the significance of such a vast sum, consider that the proposed Pentagon budget for 2018 was only $639 billion.[506] This means that the deep black budget used to fund USAPs was, back in the year 2000, *almost triple the entire Pentagon budget today*! To this startling statistic, we need to add a disquieting recent Pentagon acknowledgement that 44,000 military personnel "can't be tracked".[507] So where can the missing funds and soldiers be found? The most likely answer is in USAPs taking place within

Antarctica and elsewhere.

Consequently, while the U.S. and other nations conduct many open source science projects in Antarctica, any of these could be covers for USAPs, but without the mainstream scientists involved having any knowledge of this situation. This is certainly the conclusion that can be drawn from the whistleblower testimony of the former US Navy engineer, Brian, who was warned off by the National Security Agency from revealing anything about the events he witnessed in Antarctica, which firmly points to highly classified projects being conducted there.

According to information revealed so far in this book, there are a great number of mysterious events occurring across the ice bound continent, at least some of which are part of highly classified projects. Most disturbing is the possibility that a number of these classified projects may be illegal under the Antarctic Treaty, as well as violating other international and domestic laws. This is certainly what William Tompkins and Corey Goode claim is occurring, with forced labor being used in hidden Antarctic bases long ago established by Nazi Germany. Therefore, ascertaining what companies are conducting activities in Antarctica will aid greatly in determining if USAPs are taking place there, and whether these programs are violating the Antarctic Treaty.

Major U.S. Corporate Contractors Begin Operations in Antarctica

Soon after Operation Deep Freeze established the first permanent bases in early 1956, the US Navy took on the role of providing logistical support for the entire U.S. Antarctic program. Two years later, the National Science Foundation (NSF) started to award contracts to private companies in support of science and non-military projects at Antarctica's new U.S. stations. In 1958, the NSF began running the U.S. Antarctic program, which was the umbrella for all scientific and exploratory research on the continent. ITT was among the first companies funded by the NSF and it established a subsidiary in 1958, ITT Antarctic Services. Based out of Colorado Springs, it was ideally located to work on Unacknowledged Special Access Programs that were part of the US Air Force's evolving space program.[508] In chapter 2, it was pointed out that ITT was one of the U.S. companies that worked closely with Nazi Germany, and had established telecommunication and manufacturing subsidiaries and partnerships prior to and during WWII. Given the early establishment of ITT Antarctic Services, there is good reason to believe that as a result of the 1955 agreement between the Fourth Reich and the Eisenhower Administration, ITT quickly resumed work with its former German corporate subsidiaries and partners that were secretly operating in Antarctica.

It is important to understand how the Germans handed over the lion's share of responsibility for researching the ancient artifacts discovered in Antarctica to the U.S. military-industrial complex. This occurred after agreements had finally been reached with the Eisenhower administration. Regarding this history of events, Goode writes:

> So after the Germans had discovered this, they started to, not really excavate for archaeological reasons, they started to clear out areas and begin to use them that they found. They found in cave areas ancient civilization artifacts, and they just moved them out and moved their military groups in and created bases. So they weren't real interested in the archaeology in the beginning, which sounds not very much like the Nazis since they had archaeologists combing, you know … combing Asia looking for certain … It seemed mission was to hurry up and build this [Antarctic] base. That was the primary job … So they had been

looking into it, doing some digs here and there, over the decades leading up to where they ended up working with the Americans, the industrial complex, military-industrial complex. They were doing excavations all throughout the decades going all the way back, you know, from the '50s and '60s.[509]

Similarly, Dr. Pete Peterson says that the U.S. discovery and excavations of the three spacecraft found in Antarctica date to the 1950's, however, the Germans made the discovery first.[510]

Given ITT's historically close relationship with German corporations, the founding of its Antarctic subsidiary in 1958, ITT's expertise in advanced communications technologies, and finally, its close relationship with Reptilians according to Bill Tompkins, it's almost certain that ITT was heavily involved in the excavations of the extraterrestrial civilization then underway in Antarctica.[511]

The role of private contractors steadily increased throughout the 1960's and 1970's as projects grew and corporate support became increasingly necessary.[512] In 1972, the Antarctic Support Contract was established by the National Science Foundation and awarded to a sole major contractor to administer all of the science and non-military projects run out of U.S. Antarctic stations. Defense contractors that were awarded the Antarctic Support Contract could bring in additional expertise and personnel to assist in the secret excavations that were occurring under the lead companies, such as ITT.

In 1972, Holmes & Narver became the first corporate contractor for the NSF Antarctic Support Contract, which covered support services at all three permanent, year-round, U.S. bases – McMurdo Station, Amundsen-Scott South Pole Station and Palmer Station.[513] Notably, Holmes and Narver was one of the defense contractors that worked on the Manhattan Project.[514]

Over the successive decades, Holmes & Narver partnered with major aerospace and engineering companies such as

Raytheon and EG&G in bidding for the Antarctic Support Contract.[515] In 2000, Holmes and Narver was absorbed by the Engineering Industrial giant, AECOM.[516] Given Holmes & Narver's involvement with the 1940's Manhattan Project to develop the atomic bomb, it's fair to conclude that this corporation was very familiar with the security procedures governing Unacknowledged Special Access Programs. This would have encompassed the need to keep secret the presence of an active German space program deep below the Antarctic ice, which the U.S. military-industrial complex was cooperating with in a secret agreement.

In 1980, ITT was awarded the next Antarctic Support Contract through its subsidiary, ITT Antarctic Services.[517] A report describing ITT's support activities in the 1980-81 seasons says that it helped 77 science projects during the summer period, and five projects during the winter.[518] Holding the 1980 Antarctic Support Contract, ITT was perfectly positioned to step up its covert activities by awarding sub-contracts to any companies that could help in the secret excavations. Furthermore, ITT could ensure that U.S. Antarctic scientific operations would not negatively impact on the operational secrecy of the German Antarctic operations that worked closely with major U.S. defense contractors.

In 1990, Holmes and Narver partnered with EG&G to form Antarctic Support Associates to win the NSF's Antarctic Support Contract. [519] Formed in 1947, EG&G was a major defense contractor and worked on a range of highly classified nuclear related projects. Its "Special Projects" division was the operator of the infamous JANET Terminal at Las Vegas, which transported employees to Area 51, Nevada, where numerous USAPs were conducted in strict secrecy.[520] Indeed, EG&G was the company that hired Bob Lazar to work at the notorious Papoose Lake facility at Area 51 where he witnessed nine types of flying saucer craft.[521] Well versed in managing USAPs, EG&G was a logical partner to oversee similar programs in Antarctica, using scientific projects as cover programs, as required by law. [522]

In 1999, Raytheon was awarded the Antarctic Support

Contract by the National Science Foundation. Raytheon had long worked closely with the US Air Force, which had now gained control over Antarctic support operations from the US Navy. Raytheon's Polar Services website described its main function as:

> Raytheon Polar Services exists specifically to meet the needs of the National Science Foundation (NSF) Office of Polar Programs. The main function of RPSC is to provide support to the United States Antarctic Program ... which is dedicated to sustaining the Antarctic environment and funding scientists who conduct research in Antarctica.[523]

Raytheon was first established in 1922, and has grown into a giant in the aerospace industry. In 2015, it became the 3rd largest defense contractor in the U.S., and was fifth largest in the world.[524] Here is how Raytheon describes itself on its website:

> Raytheon Company is a technology and innovation leader specializing in defense, civil government and cybersecurity solutions. Founded in 1922, Raytheon provides state-of-the-art electronics, mission systems integration, capabilities in C5I (command, control, communications, computing, cyber and intelligence), sensing, effects and mission support services. Raytheon is headquartered in Waltham, Massachusetts.[525]

Raytheon's website further adds that in 2017, it employed 63,000 people and grossed over $24 billion in sales. Raytheon regularly advertises jobs requiring employees to have security clearances for Special Access Programs.[526]

Audits of Raytheon Polar Services over the calendar years of 2000 to 2004 indicated that it was indeed conducting covert operations in Antarctica, as claimed by Richard Hoagland and

Mike Bara. [527] The funding of covert operations was achieved through a black budget that Raytheon had set up, which is evidenced by irregularities found through the audits:

> In our opinion, the indirect cost and other direct cost system and related internal controls of Raytheon Polar Services Company are inadequate. Our examination identified certain significant deficiencies in the design or operation of the internal control structure. In our opinion, these deficiencies could adversely affect the organization's ability to record, process, summarize, and report indirect and ODC costs in a manner that is consistent with applicable government contract laws and regulations.[528]

Despite auditing irregularities being reported as early as 2004, Raytheon continued to be awarded contracts to continue its support services in Antarctica, as exemplified in its April 5, 2010 Press Release announcing a one year extension of operations:

> Raytheon Polar Services has earned excellent performance ratings since 2000 as the prime contractor to the U.S. Antarctic Program for NSF. The extension will take the contract to March 31, 2011.
>
> "We appreciate the opportunity to continue support for the valuable scientific research underway at the bottom of the world. This is a keystone program for us," said Sam Feola, Raytheon Polar Services program director.
>
> Raytheon Polar Services employs about 350 full-time staff and hires approximately 1,400 contract employees from its offices in Centennial, Colo...[529]

On September 20, 2010, it was announced that Raytheon was given another one year extension by the National Science Foundation despite auditing irregularities and controversy surrounding its policies.[530] Raytheon's contract ended on March 30, 2012, after its 10 year contract was given two unprecedented one year extensions.

On Dec 28, 2011, it was announced that Lockheed Martin had been awarded the National Science Foundation's $2 billion Antarctic Support Contract. [531] Lockheed Martin took over Antarctic support operations on April 1, 2012, for an initial contract period of 4.5 years, which could be extended to 8.5 years. Lockheed Martin is the world's largest defense contractor with 97,000 employees worldwide, and over $45 billion in government contracts awarded in 2009.[532]

One of the two predecessors of Lockheed Martin, the Lockheed Corporation, was among the U.S. companies that William Tompkins says he delivered briefing packets to, containing information about German flying saucer developments during World War II. Tompkins took the packages to Lockheed's Advanced Development Projects (aka Skunkworks), which was created on June 1943 by the legendary Clarence "Kelly" Johnson.[533] From December 1942 to January 1946, Tompkins says that he visited Lockheed's advanced research and development facilities several times.[534] Significantly, Tompkins states that the briefing packages contained information about the secret Antarctic operations being conducted by the Germans. [535]Furthermore, in response to a question about which U.S. corporations cooperated most closely with the Antarctica based Germans, Tompkins' reply included the Lockheed Corporation.[536]

Lockheed's Skunkworks has long been associated the US Air Force and the Central Intelligence Agency in building advanced aircraft at Area 51 that were highly classified Unacknowledged Special Access Programs.[537] Indeed, declassified records show that Kelly Johnson was part of the original team that picked Area 51 as a suitable location for conducting USAPs such as the U-2 spy plane and the SR-71 Blackbird.[538]   In his autobiography, Ben Rich describes how Lockheed's Skunk Works worked predominately on highly classified US Air Force contracts, and how difficult it was to work with the Navy.[539]

Subsequently, like its predecessors, Raytheon and EG&G (now URS), Lockheed Martin had a long history in conducting USAPs. It was more than capable of managing highly classified Antarctic operations, through cover programs, which would be funded through the Antarctic Support Contract.

On August 22, 2016, the National Science Foundation announced the transfer of the Lockheed Martin contract to the newly reconstituted company, Leidos, as a result of a corporate merger:

> Leidos Holdings, Inc. will hold the National Science Foundation's contract for support of the U.S. Antarctic Program (USAP). The change follows the merger between Leidos and Lockheed Martin's Information Systems & Global Solutions business segment as of Aug. 16. Lockheed Martin has held the contract since NSF made the award in December 2011.[540]

Leidos Holdings was formed on September 27, 2013 as a spin off from Science Applications International Corporation (SAIC), which had divided itself in two. SAIC had been established in 1969 by a former US Navy officer and leading nuclear physicist, Dr. J. Robert Beyster.[541] He worked closely with the National Security Agency and the military intelligence community in gaining contracts for his new company, which built significantly on his expertise in the nuclear energy industry.

After SAIC's 2013 corporate split, Leidos continued to specialize in SAIC's former work in national security and defense contracts, while the rump SAIC would retain the parent company name, and focus on information technology.[542] In a $5 billion business deal, Leidos purchased Lockheed's Information Systems & Global Solutions division, and was now in charge of the Antarctic Support Contract, which runs up to 2025.[543] In its former incarnation as SAIC, Leidos had acquired much experience in running Unacknowledged Special Access Programs. SAIC was involved in the development of antigravity craft for the US Air Force according to the late whistleblower, Colonel Steve Wilson.[544] Furthermore, SAIC had on its Board of Directors Admiral Bobby Ray Inman who, according to William Tompkins, was a key figure in developing designs for future antigravity fleets of Navy spacecraft.

Admiral Bobby Ray Inman & the NSA SAIC/Leidos Connection to Antarctica

Tompkins, who worked at Douglas Aircraft Company's Advanced Design think tank from 1950 to 1963, described how Inman would transmit unsolicited spacecraft design proposals developed at Douglas to a very high level Navy intelligence group of officers. In his book, *Selected by Extraterrestrials*, Tompkins referred on several occasions to Admiral Inman as a key figure in the development of unsolicited bids for the Navy to develop antigravity spacecraft.[545] In the following passage, Tompkins recalled what he was told about Inman's role by his boss at Douglas Aircraft Company, Elmer Wheaton:

> Now, Bill, I want you to understand that every time
> – since your first top secret disseminator
> assignment in Naval Intelligence – you were
> requested to visit or work in a classified military
> facility and you were never ever denied access.
> Even admirals commanding Battle Groups in
> combat were denied access that you can just walk

right into. The boy with the photographic memory was a lot more knowledgeable about Naval space combat than them.

You have been tracked by both extraterrestrials and an elite Naval intelligence-gathering operation since Commander Perry reviewed your Naval ship model collection documentation in 1940. You were never informed of your association with the alien alliance. Nor your highest security classifications as a contactee monitored by a small core of senior Naval officers probably chaired by Bobby Ray Inman.[546]

The editor of *Selected by Extraterrestrials*, Dr. Robert Wood, who also worked at Douglas during Tompkins' employment there, inserted the following note which elaborated on Inman's role in what occurred at the secret Douglas Think Tank:

(Note from Editor Wood. My interpretation of these remarks and their chronology is this: Elmer Wheaton had contact with the UFO-cleared group in the Navy, which he referred to as "Forrestal's people" as the ones who knew about the UFO issues. One of the new young Navy officers who was cleared for the UFO topic appears to have been Bobby Ray Inman, and his inside knowledge of the UFO problem may well have been the special link to his subsequent highly successful career. Apparently Bobby Ray was the main person interacting with the Wheaton think tank at the time of this conversation. Since Bill Tompkins' time in this vault spans several years, it is not really clear that this conversation occurred in 1952 or perhaps a year or so later.)[547]

I spoke by phone with Admiral Inman on December 1, 2016, and he denied having any involvement during the 1950's and 1960's with anything occurring at Douglas Aircraft Company.[548] However, he did point out the chronology of his Naval career, including his later involvement with the NSA and SAIC, which is significant given Tompkins' claims. Based on his historical role with the intelligence community and subsequent work on Unacknowledged Special Access Programs (USAPs) with multiple corporations, Tompkins' claims appear very plausible. This is despite Inman's denial of having any association with Douglas, which he was legally required to do as part of the standard security procedures in place for USAPs.

Inman said that after attending a postgraduate program in Naval Intelligence in Washington, DC in 1958, he stayed on at the Pentagon as an intelligence briefer until 1960, and next served on a destroyer in the Atlantic. In October 1961, he worked in a Naval office housed within the NSA until 1965. He then transferred to Hawaii (Pacific Command) where he headed Naval Intelligence until July 1967. After his Hawaii assignment, he became the Intelligence officer for the 7th Fleet, from May 1969 to August 1971.

In January 1974, Inman received his first star (Rear Admiral lower class) and became both Director of Naval Intelligence and headed the National Underwater Reconnaissance Office, leaving these positions in July 1976. He then moved over to the Defense Intelligence Agency where he was Vice Director up to 1977, and moved again to become Director of the National Security Agency up to 1981. He finally became Deputy Director of the CIA, from February 1981 to June 1982.

Admiral Inman acknowledged in our phone conversation that during his naval career, many companies had worked for him on various contracts awarded to them by the Navy. Regarding the bidding process for naval contracts, he said that he was involved in setting forth the requirements for a corporate bid, and then reviewing the work done by the corporations in completing a bid.

Among the companies that had bid successfully for Navy contracts was Science Applications International Corporation, whose Board of Directors Inman joined after his official retirement from government and military service in 1982.[549] Inman would go on to serve for 21 years on SAIC's Board of Directors, until retiring from it on October 1, 2003.[550] This is where it is helpful to understand the close relationship between the NSA and SAIC, which had become a revolving door where senior NSA officials would become SAIC executives, and then be moved back into more senior positions within the NSA. In his book, *The Shadow Factory*, James Bamford described how the revolving door between the NSA and SAIC operated:

> After first installing the former NSA director Bob Inman on its board, SAIC then hired top agency official William B. Black Jr. as a vice president following his retirement in 1997. Then Mike Hayden hired him back to be the agency's deputy director in 2000. Two years later SAIC won the $280 million Trailblazer contract to help develop the agency's next generation eavesdropping architecture, which Black managed. Another official spinning back and forth between the company and the agency was Samuel S. Visner. From 1997 to 2001 he was SAIC's vice president for corporate development. He then moved to Fort Meade as the NSA's chief of signals intelligence programs and two years later returned as a senior vice president and director of SAIC's strategic planning and business development with the company's intelligence group.[551]

SAIC has long been considered to be one of the most "spook infested" companies involved in highly classified programs run by the U.S. military intelligence community.[552] Given what we know of the NSA's role in silencing the Navy whistleblower Brian from speaking further about his Antarctic experience in 2016, it's significant that Leidos (formerly SAIC) was awarded the Antarctic Support Contract around the same time Brian was being intimidated into silence. This is a reflection of the historically close relationship between Leidos, the NSA and the US Navy.

When Inman was on the SAIC Board of Directors (1983-84), the corporation was given a contract to conduct a meteorite study in Antarctica to better understand conditions for future Mars operations. Here is how *The Antarctic Sun* described the SAIC project:

> Like Antarctica, Mars is a polar desert. Without water eroding the surface, any weathering is from wind or sun, just like the Dry Valleys, said Dean Eppler, a NASA consultant with Science Applications International Corp., who studied the weather patterns in the Valleys in 1983 and 1984 in order to better understand Mars. The Dry Valleys look almost identical to photos the Viking lander sent back from Mars "right down to the rocks that are there," Eppler said.[553]

It's almost certain that the SAIC study was a cover program for a more classified project it was conducting, under the auspices of NASA and ITT Antarctic Services which held the Antarctic Support Contract at the time.[554]

At this point, it is important to recall what Joseph McMoneagle reported during his remote viewing session; he witnessed a Martian exodus to a volcanic area on another world, presumably Antarctica.[555] The session was conducted in May 1984, the same period when SAIC was involved in Mars and Antarctica research. In a 2004 lecture, McMoneagle said he was in his final year of Project Stargate when his trainer, F. Holmes "Skip" Atwater, brought in the Mars coordinates. Atwater, discussing the 1984 remote viewing session, said that the coordinates had been provided by Dr. Hal Puthoff, the Director of the Stanford Research Institute's remote viewing program, Project Stargate, which formally began in 1972 with CIA funding as Project Scanate ("scan by coordinate").[556]

U.S. military intelligence involvement in Project Stargate began in 1977 as the Gondola Wish program, which was undertaken by Army Intelligence and the Defense Intelligence Agency (DIA) at Fort Meade, Maryland.[557] It's worth noting that Inman was Vice Director of the DIA from 1976 to 1977, so it's possible that he was aware of the efficacy of remote viewing and the formal partnership discussions then underway with SAIC.

It has been documented that SAIC was one of the organizations actively studying and using remote viewing as an intelligence gathering tool.[558] Either during his time at the DIA and/or as Director of the NSA (1977-81), and/or Deputy Director of the CIA (1981-82), Inman was certainly briefed on remote viewing and its purported accuracy. After joining the SAIC Board of Directors upon his 1982 retirement, Inman therefore likely recommended the use of remote viewing in SAIC's classified work.

Given this situation, it's highly probable that SAIC was involved at some level with McMoneagle's remote viewing session, or at the very least, learned of its results in order to better understand what was being excavated in Antarctica. It's reasonable to conclude the SAIC Mars meteorite study was a cover program for SAIC's participation in the classified study of the remnants of the Martian civilization buried deep under the Antarctic ice. Inman's involvement, through his position on the Board of Directors, suggests that SAIC was doing far more than merely searching for Martian meteorites on Antarctica's icy surface.

Understanding the history of SAIC in Antarctica, along with Inman's participation, is important background information for understanding the 2013 decision to split the company in two. By divesting itself of its non-national security components through the 2013 corporate separation, Leidos became better suited to conducting its more traditional national security programs under the required operational security protocols. Consequently, there is reason to believe that Leidos was encouraged or allowed to take over the Antarctic Support Contract, so it could become more involved in managing the Unacknowledged Special Access Programs occurring there.

Perhaps even more significant is Liedos' historical affiliation with US Navy interests, as opposed to Lockheed Martin's closer historical relationship with the US Air Force. Therefore, the shift of the Antarctic Support Contract from Lockheed Martin to Leidos may be a direct result of the US Navy's desire to reassert its earlier authority over highly classified programs in Antarctica as current events reach a critical level there.

Use of Slave Labor in R & D Antarctic Project by Transnational Corporations

On July 31, 2017, Corey Goode forwarded the following Skype message to me discussing a briefing he had just received from "Gonzales", the alleged US Navy Lt. Commander who has been Goode's primary contact with a "Secret Space Program Alliance" since late 2014:

> Spoke to Gonzales more in depth that I have before. He told me he was once assigned to a few of the facilities that were built during the 60's in Antarctica. He wanted me to know why the Cabal was reacting so harshly to us disclosing the info Sigmund gave us about the R&D Facilities under the ice. It is MUCH worse than I was first led to believe. There are literally thousands of abducted

humans down there that are being used in experiments.[559]

In his Skype communication, Goode also elaborated on the subject of the captive humans used in these secret Research and Development (R&D) facilities hidden deep below the Antarctic ice sheets:

> Gonzales said this was one of the parts of his service that was the hardest to come to terms with. He said that the people working there would dehumanize the subjects to be able to do the work. People that were not psychopaths would behave like one to be able to make it through doing the work ... Lots of medical and genetic research. They test these people to death in many horrible ways ... I was told that the number went from as low as 10K to as many as 40K people being used until they are dead in these facilities that honeycomb one large area alone ... Mostly from the human slave trade.[560]

In his response to a question I asked about what kind of human rights violations and experiments were occurring at secret Antarctic bases, Goode replied: "These R&D Bases do Nuclear, Biological and Radiation Experiments on Humans. There are a number of human cloning operations going on in these and other bases."[561]

This testimony by Goode may come as no surprise given what was discussed earlier in chapter 6 regarding the slave labor practices underway in Antarctica; first implemented by the Germans and later adopted by the U.S. military-industrial complex. Next, Goode went on to explain the prominent role of defense contractors in the human rights abuses happening in Antarctica:

> Super sick stuff and just about all of the research
> being done down there by defense contractors is
> beyond illegal/unethical. Scarrry shit! He
> [Gonzales] did say if we stopped talking about the
> MIC Space Port and the R&D Facilities that our lives
> would return to normal. He said it just depends on
> what type of disclosures we are willing to make.[562]

Importantly, Goode is referring here to the defense contractors working on highly classified projects in Antarctica conducting the illegal experiments and abuses. It's worth pointing out that such systematic abuses could occur in Unacknowledged Special Access Programs (USAPs), because they prevent the necessary independent oversight and scrutiny that would typically lead to third parties preventing such abuses from occurring. In the case of a waived USAP, the heads of a Congressional committee are only given a verbal briefing and have to publicly deny the existence of such programs. The U.S. Congress, therefore, fails dismally in fulfilling any kind of effective oversight role when it comes to USAPs.

It is critical to emphasize here that all of the companies that have been awarded the Antarctic Support Contract have been defense contractors with vast amounts of experience in managing USAPs. Holmes and Narver (now AECOM), ITT (now Exelis),[563] Raytheon, EG&G (now URS), Lockheed Martin, and most recently Leidos (formerly SAIC), are all therefore implicated in the claims that systematic human rights abuses have been occurring for decades deep under the Antarctic ice sheets by U.S. defense contractors.

In addition, according to Tompkins, companies such as Northrup Grumman and Boeing are also implicated due to their supply and delivery of personnel and resources to the Antarctic German colony for building new generations of spacecraft. In an April 2016 interview, I asked Tompkins whether Northrup Grumman and Boeing were building spacecraft carriers for the US

Navy or for the Germans in Antarctica? He responded:

> That has to be both, Michael, because we send a whole bunch of people down there [Antarctica] from commercial companies here. And we have programs that we're supposed to engineer and design. And then we have our own people back here who try to figure out "Okay, how can we take advantage of what we learned down there to help us make money here?" And so, then you'll ask the question, "Okay, we can make money here from what we learned down there and then also there may be something that we could do commercially out there."[564]

It's important to note that none of the previously mentioned U.S. companies can be accused of starting the abhorrent slave labor practices in Antarctica; they were first established by the German breakaway group. This took place because the German companies had acquired a history of using slave labor as a result of Nazi state policy, and became very proficient in using it under Hans Kammler in their advanced weapons projects to maintain secrecy. Kammler demonstrated during World War II that secrecy was best maintained using slave labor, rather than indigenous German workers. Therefore, in the joint Antarctic projects the use of slave labor, rather than American workers, was very likely a "policy requirement" imposed upon U.S. defense contractors.

By virtue of the secret agreement reached under Eisenhower, U.S. companies were infiltrated to varying degrees, and coopted into taking the existing German Antarctic program to a higher level of industrial production. Regarding the extent of the infiltration of the U.S. military-industrial complex, Bill Tompkins stated: "The agreement that Eisenhower reached basically was that these Operation Paperclip scientists would be allowed to get to the top of the U.S. military-industrial complex."[565]

From the mid 1950's onward, the German Space Program's research and development became a joint venture with the U.S. military-industrial complex. Acceptance of the Germans' previously established practice of slave labor and human experimentation was the price to be paid for full collaboration in projects that were deemed essential to the development of future U.S. space programs.

Challenges in Investigating Claims of Slave Labor in Antarctica

When attempting to confirm the validity of Goode and Tompkins' testimony about defense contractors being involved in advanced technology programs where slave labor and other abuses have occurred, a number of factors arise as challenges. First, the U.S. Freedom of Information Act (5 U.S.C. § 552) only partially applies to private entities such as defense contractors. Only documents submitted by contractors to the U.S. military or government are subject to Freedom of Information requests. Furthermore, public requests for this category of documents are subject to an exemption designed to protect trade secrets:

> Exemption 4 protects two categories of information: (1) trade secrets and (2) commercial or financial information obtained from a person that is privileged or confidential. Exemption 4 is unique in that it is designed to ensure the availability and reliability of information submitted to the government by assuring submitters that their information will remain safeguarded to prevent competitive disadvantage.[566]

It's highly doubtful that U.S. defense contractors involved in the use of slave labor would directly report on such practices in documentary correspondence with the contracting agency from the Department of Defense, let alone the National Science Foundation. It's more likely that such abhorrent practices would be cloaked behind the opaque operating procedures used in a USAP.

The second challenge involves whistleblowers. There is effectively no way for a corporate employee to potentially blow the whistle on such abuses taking place without disclosing key details of the relevant USAP, which is illegal and carries severe penalties. All military and corporate personnel involved in a USAP sign a binding Non-Disclosure Agreement which stipulates the penalties for unauthorized disclosure and possession of any evidence related to the USAPs existence. Therefore, physically possessing evidence or giving testimony of such abuses in a USAP would constitute a violation of the Non-Disclosure Agreement signed by the would be whistleblower, and could lead to his/her arrest and incarceration without the public ever learning the truth about the alleged abuses.

Furthermore, the 2013 National Defense Authorization Act only provides whistleblower protection to contractors or subcontractors who report fraud, waste or abuse occurring in the Department of Defense contract. Neither the National Defense Authorization Act, nor the Whistleblower Protection Enhancement Act of 2012, provide whistleblower protection to intelligence community contractors. [567] This means that contracts awarded by intelligence agencies such as the National Security Agency do not provide whistleblower protection for those wanting to report abuse if internal channels prove non-responsive.

Cases, like the one involving Brian, the Navy flight engineer, are rare. He did not sign a non-disclosure agreement, but was debriefed after certain incidents took place not to reveal what he had seen. Nevertheless, he courageously blew the whistle on what he had witnessed during his 14 year career in Antarctica (1983-97). Even though Brian did not possess any evidence confirming what he had actually seen, when he attempted to tell more details about the mysterious events he had seen, he was threatened into silence by the NSA. Such threat tactics are consistent with how USAP's generally maintain operational security, thus suggesting in this case that such programs were indeed occurring in Antarctica as Tompkins and Goode have claimed.

When Congressman Nicholas Lehman and Dr. Rita Colwell, members of a 2002 Congressional Delegation to Antarctica, [568] were questioned about the large hole at the South Pole that Brian had witnessed, neither admitted to any knowledge of it. [569] In response to further questions about classified programs in Antarctica, neither acknowledged being told anything about the existence of such programs.

Third, the annual reports submitted by defense contractors that have held the Antarctic Support Contract reveal few details about the classified programs the corporations are actually involved in. The reports primarily allude to the business risk of running such programs in terms of profit margins, rather than possible ethical risks. For example, in its 2016 Annual Report, Leidos acknowledged its reliance on classified programs for its revenue, and also the risk this poses for investors who cannot find out what is occurring:

> We derive a portion of our revenues from programs with the U.S. Government that are subject to security restrictions (classified programs), which preclude the dissemination of information that is classified for national security purposes. We are limited in our ability to provide information about these classified programs, their risks or any disputes or claims relating to such programs. As a result, investors have less insight

into our classified programs than our other businesses and therefore less ability to fully evaluate the risks related to our classified business.[570]

Similarly, in its 2017 Annual Report, Exelis/Harris said:

Although classified programs generally are not discussed in this Report, the operating results relating to classified programs are included in our Consolidated Financial Statements. We believe that the business risks associated with our classified programs do not differ materially from the business risks associated with our other U.S. Government programs.[571]

Fourth, these primary defense contractors award sub-contracts to other companies that provide various support functions to both the open science program and the highly classified programs (USAPs) occurring in Antarctica. Thus, the primary contractor can hand over full control to a subcontractor for a particular USAP, which is then funded off the books by black budget funds funneled through the CIA to the contractor or subcontractor.[572]

Finally, there are the respective roles played by NSA, the CIA and NASA in the USAPs underway in Antarctica, which have been awarded to different defense contractors. The NSA, as mentioned earlier in the case of Brian, was involved in maintaining operational security for USAPs in Antarctica. Similarly, according to Hoagland and Bara, the NSA was involved in the classified programs set up to explore the Lake Vostok magnetic anomaly. Consequently, the NSA has been institutionally involved in maintaining secrecy over Antarctica's USAPs, and the illegal research and development occurring there.

The CIA, under the leadership of Allen Dulles, played a

critical role in the negotiations that led to the Eisenhower administration's agreement with the German Antarctic Space Program. Furthermore, the CIA's role as the U.S. entity responsible for funneling black budget funds through different U.S. agencies, would give it significant authority in USAP operations and how these are managed.[573]

In the case of NASA, according to Tompkins, McClelland and Goode, it was thoroughly penetrated by German scientists who had joined the public space program, but were fifth columnists for the German Antarctic Space Program. NASA is involved in a number of open source science programs in Antarctica, such as the meteorite collection program, which can be used to cover its involvement in USAPs deep under the ice sheets.[574]

Despite the challenges in investigating illegal research and development in Antarctica and the alleged use of slave labor there, the evidence available so far indicates this is occurring and cannot be ignored. Thus, strategy for dealing with such illegality needs to be explored, as well as how to facilitate full disclosure of the corporate-run space programs and Antarctica's hidden history.

CHAPTER 16

Antarctica's Suppressed History & Full Disclosure

Historic material examined in this book shows that immediately after the end of World War I, a powerful German secret society (Thule *Gesellshaft*) was covertly supported by German Navy Intelligence officials who sought to promote a number of nationalist projects. These projects were aimed at preserving Germany's unity and restoring its place amongst the world's leading nations as quickly as possible, despite the restrictions imposed by the Treaty of Versailles. Amid these nationalist projects in the early 1920's came the support for a female mystic, Maria Orsic, who prominent Thule Society members believed was psychically communicating with beings from Alderbaran and/or the Inner Earth. Orsic's psychic information appeared genuine after providing the secrets of advanced technologies that would make space travel possible, and also offered to transform German society by giving citizens first-hand knowledge of a mysterious universal force, called Vril.

The collaboration between Orsic and the Thule Society established the foundation for the evolution of two distinct space programs in the subsequent decades. One was led by Orsic out of secret bases in Antarctica and South America, and taught a cosmic philosophy of peace and cooperation, which gave rise to the Space Brothers phenomenon in the 1950's. The other German space program had a much darker vision based on unrestrained weapons development, the exploitation of slave labor, and imperial conquest – all policies explicitly sanctioned by another Thule protégé, Adolf Hitler.

What historical records clearly show is that the Thule Society, German industrialists and German Navy officials all supported Hitler's rise to power due to his unfettered nationalist fervor and desire to restore Germany's military power. These records indisputably show that a substantial portion of the funds and resources necessary for rebuilding Germany and its military came from U.S. companies. Prominent U.S. corporate and political figures such as Henry Ford, John Foster and Allen Dulles, John D. Rockefeller, Prescott Bush and many others aided and abetted Hitler's rise to power, and the subsequent rearming of Germany through international banking and investments that poured into Nazi Germany.

This continued well into World War II, during which time President Roosevelt granted exemptions to major U.S. companies, enabling continued trade with Nazi Germany despite the "Trading with the Enemy Act". This was also done while turning a blind eye to the glaring fact that Nazi Germany was systematically using slave labor in its industrial production facilities, some of which had been built with U.S. financial resources. As the end of the war approached, deals were secretly reached which allowed the Nazis to transfer substantial financial resources out of the country to later be used in a secret plan developed by Martin Bormann to fund the rise of a Fourth Reich. This unconscionable process once again promised significant profits for cooperating U.S. corporations.

After the war, many of these same U.S. corporate and political figures led the secret negotiations that brought in the remnants of Nazi Germany's most advanced weapons projects, under Operation Paperclip, into the U.S. All of this is well known and established historical fact. Not so well known is the extent to which the Fourth Reich infiltrated and compromised the U.S. military-industrial complex. This took place because many U.S. and German industrialists belonged to a worldwide fraternity of societal elites, whose allegiance to one another transcended national affiliations. This made the establishment of multiple secret space programs possible in the remote regions of South America and Antarctica.

The first of these was the Vril Society program run by Orsic, which maintained its benign "spiritual" vision and operations, as evidenced by the 1950's and 1960's Space Brothers phenomenon. Orsic and other German astronauts pretended to be extraterrestrials while meeting with many ordinary civilians in the U.S. and other countries. Orsic's Vril Space Program hoped to seed ideas of cosmic unity and higher consciousness to help start a global revolution. They were helped by human-looking "Nordic" extraterrestrials and Inner Earth beings who shared these benevolent "service-to-other" goals.

At the same time, other German Secret Society members cooperated with Reptilian extraterrestrials in setting up an aggressive imperialistic space program. The initial goal was world conquest, whereby the Earth's industrial resources would be devoted to building a mercenary space force that would fight alongside the Reptilian's fleets in deep space. After the failure of Hitler's military campaigns, a more covert plan was initiated using the advanced flying saucer technologies developed in Antarctica by the Fourth Reich. After the defeat of Admiral Byrd's 1947 "Operation Highjump" military expedition, the Fourth Reich's spacecraft proceeded to boldly fly over U.S. territories regularly, culminating in the highly dramatic 1952 Washington Flyover. This critical event forced the Truman, and later Eisenhower, administrations to embark on secret peace negotiations which effectively amounted to a negotiated surrender by 1955. The secret agreements with the Fourth Reich were kept a closely guarded "national security" secret from the American and world public, while the U.S. military precariously hoped to bridge the technological gap.

The subsequent flow of American industrial resources into Antarctica led to a significant expansion of the Fourth Reich's imperial space program. It also enabled the establishment of yet another space program; founded in a partnership between the Fourth Reich and the U.S. military-industrial complex. This ultimately led to the creation of a transnational corporate space program called the Interplanetary Corporate Conglomerate. According to Corey Goode, this corporate space program became operational in the late 1980's, after the launch of the US Navy's "Solar Warden" Space Program.[575] This transnational collaboration further led to several major U.S. corporations cooperating in projects that exploited slave labor and used human subjects in large industrial scale operations hidden deep below the Antarctic ice sheets.

The establishment of parallel German space programs in Antarctica during the 1940's and 1950's, by secret societies and leading industrialists, comprised of a highly spiritual ethical faction (Orsic's "Space Brothers" Program) and an aggressive imperialistic faction (Fourth Reich) who ultimately established a partnership with the U.S. military-industrial complex, are relatively recent chapters in Antarctica's suppressed history. The presence of an ancient Martian colony, perfectly preserved deep under the Antarctic ice since the Great Flood, and its secret excavation since 2002, represent a far older chapter of Antarctica's suppressed history which needs to be revealed in a genuine "full disclosure" initiative.

Most significant is the likelihood that humanity will eventually learn of the existence of "Pre-Adamites" or "Ancient Martians", who may still be alive in stasis chambers located in one or more extraterrestrial motherships. The possibility that these beings are the imprisoned "fallen angels" described in the Book of Enoch will have enormous implications for religious scholars and

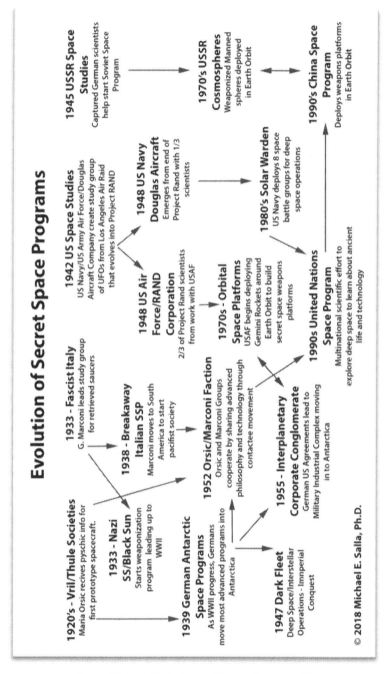

Figure 68. Evolution of Secret Space Programs Chart

humanity in general. Predictably, this will lead to great debate over the wisdom of awakening, and/or cooperating with, such beings in the future.

The first issue that arises, however, is whether the general public will be informed in any way about the extraterrestrial element within the Antarctica excavations. Corey Goode describes a "limited disclosure" plan, which involves sanitizing the Antarctica archeological sites of any remains having to do with extraterrestrial life:

> Now, another interesting note, we do have these archaeologists and employees of various universities that are down there excavating and documenting all of this, but what they have done, they being the Cabal, I guess you'll say, they have used these large electromagnetic submarines that I discussed earlier to take a lot of archaeological items that they had found in other digs that they were keeping suppressed from humanity, they had them in huge warehouses. [576]

> They were taking some of these artifacts down to Antarctica and seeding them. And this one large dig that these archaeologists are going to make public. They are also removing any body that does not look human. [577]

The next step in this limited disclosure plan, according to Goode, is to slowly reveal in a many decades long process, the existence of secret space programs:

> So they're planning on giving us a sanitized disclosure, and then over time they will disclose the Military-Industrial Complex Secret Space Program.

> And after they do that, they'll say, "Oh, by the way, yeah, we've got this fairly advanced Secret Space Program, and while we've been out to other planets, we've found very similar ruins as we've found in Antarctica.
>
> So they're going to try to trickle the information down over decades, and not immediately tell us about the ships they found and the high technology and non-humans. [578]

Clearly, such "limited disclosure" is unacceptable to all familiar with the true history of Antarctica as discussed in this book. What will prove a critical issue for "full disclosure" to take place is uncovering the precise nature of the relationship between these giant Pre-Adamites and the world's ruling elite. Are the latter the direct bloodline descendants of ancient hybrid Pre-Adamite rulers; described in the Sumerian Kings List and Manetho's fragments of Egyptian history as "demigods"? Also, what is their relationship with the long-lived pre-diluvial patriarchs described in the Old Testament? The policy implications are enormous as we learn that many powerful secret societies, long suspected to be the hidden rulers of the planet, may in fact be covert worshippers of "fallen angels" destined to awaken from the frozen depths of the Antarctic continent.

It's no coincidence that the Thule Society, the hidden force behind the rise of Hitler's Third Reich, believed Aryans to be the descendants of a pre-diluvial Hyperborean super race that once ruled over the Northern Hemisphere. Similar beliefs have been shared by other secret societies within the U.S., United Kingdom and other nations. Their shared beliefs facilitated a close cooperation between secret societies and industrial elites leading up to, during, and well after the end of World War II. This connection is especially significant when it comes to the evolution of the secret space program, the "Interplanetary Corporate

Conglomerate", run by transnational corporations out of the German's former Antarctic bases.

Furthermore, we need to consider the policy implications concerning agreements that German secret societies entered into with Reptilian extraterrestrial's established in Antarctica, which led to a space program that Goode has termed the "Dark Fleet". This German/Reptilian collaboration is alleged to involve interstellar conquest and a galactic slave trade, which William Tompkins and Goode believe is a major imperative to end. Just prior to Tompkins' surprising, if not suspicious, passing on August 21, 2017, he aptly summarized part of the challenge that lies ahead in learning the truth about events in Antarctica:

> Germany in World War 2 had massive underground facilities with workers that were all slaves, and even to the extent that when the decision was made before the war ended that the Germans were going to continue all of their extraterrestrial developments on UFOs, and on every weapon system that they were building, the Germans took these production facilities to Antarctica, and they also brought their war slaves with them to continue their work.

> So now there are slaves working underneath the ground, and they still are working underground today in Antarctica. But the slave business out there in the Milky Way is a big business, and this is happening today. It's not something that happened one hundred years ago. This slavery has been going on a long time, and **needs to be stopped.**[579]

Goode has stated that the Dark Fleet and the Interplanetary Corporate Conglomerate have become close allies and collaborated in a galactic slave trade out of Antarctica. Major

German corporations such as Siemans, has been directly involved in building key components for the Antarctic secret space program, and has also been implicated in this human trafficking problem. Not only are individuals being forcefully taken to work as slave labor in illegal research and development programs in Antarctica, but they are also being taken off planet into a galactic slave trade enterprise. Therefore, Antarctica has become a major spaceport for this galactic slave trade.

It's worth pointing out that this is no accident. Albert Speer exposed that a post-World War II "peace-time" global slave economy was actually planned by the Third Reich.[580] The evidence presented in this book lays out the trail of how the Fourth Reich covertly continued this egregious plan not only upon this planet, but in its deep space operations as well. The extent to which U.S. defense contractors participate in a galactic slave trade, in addition to the use of slave labor in Unacknowledged Special Access Programs, is not clear. It's likely, however, to be substantial.

Even if the scenario described by Tompkins and Goode about the use of force labor and a galactic slave trade is only partly accurate, then they have provided an answer to what has happened to millions of people, the majority of whom are children, who mysteriously disappear every year around the world.[581] The prospect of Antarctica currently being exploited for a number of highly classified projects, run by transnational corporations that use slave labor, involve human experimentation, and even the establishment of a spaceport for a galactic slave trade, is deeply disturbing.

Exposing such abusive practices, and the companies and government agencies implicated in them, is an important first step to address a great historic injustice. In this book, a number of German and U.S. companies that have been involved in Antarctica's secret space programs have been identified to varying degrees. Each company deserves serious scrutiny by stockholders, the general public and appropriate investigative bodies, to find out the full extent of each company's involvement in such alleged abuses.

The next logical step is to free the victims of such illegal practices and implement a form of restorative justice, where they, their families or communities are aided or compensated for what the victims have endured over extended periods. "Restorative Justice" is a far older model of justice, which seeks to address the injuries suffered by the victims of crime by the perpetrator directly negotiating with the victims, their families or community to mutually agree upon an outcome. Restorative Justice differs from the more common "Retributive Justice" model used in the developed world which views crimes more as a violation of state sanctioned laws, thereby requiring the punishment of perpetrators, rather than addressing the needs of victims and their families or communities. The difference between these two justice models is aptly summarized as follows by Wikipedia:

> Restorative justice is an approach to justice that personalizes the crime by having the victims and the offenders mediate a restitution agreement to the satisfaction of each, as well as involving the community. This contrasts to more punitive approaches where the main aim is retributive justice or to satisfy abstract legal principles.[582]

The final step in addressing such a great historic injustice caused by the use of forced labor, and the creation of an off-world slave trade out of Antarctica, is to put major global reforms in place that make it impossible for transnational corporations to ever again perform such practices under the cloak of trade secrets and national security. In the U.S., the security protocols in place for Unacknowledged Special Access Programs make it extremely difficult, if not impossible, for independent parties to find out what happens in such programs, especially when they are contracted out to transnational corporations. There needs to be an independent global authority with the resources and means to investigate what happens in corporate laboratories. Just as there currently exists an international regulatory body to investigate nation states suspected of trafficking in illegal nuclear weapons or biological research, there needs to a similar global authority for investigating transnational corporations for any kind of illegal research and development.

The preceding steps, along with others, will greatly aid in promoting freedom and sanctity at every level of human society, especially as our world is transformed due to incredible technological advances, along with our suppressed history emerging as the truth about Antarctica is revealed. Most importantly, these steps will help bring the integrity back into the global national security system that has been out of balance far too long. The necessity for doing so is not merely a moral argument, but an important national security requirement.

One lesson that history clearly teaches is that a political system or civilization, no matter how powerful it may appear at the time, will crumble if it is out of integrity with its citizenry. Power that is exercised without abiding by the archetypes of "Truth and Justice" will always be perceived as a form of tyranny by its citizens, thereby weakening the political system from within. History is filled with stories of failed civilizations and kingdoms that crumbled when the political elite did not abide by the imperatives of Truth and Justice. Most often, this occurred by the incursion of external forces, where an invading force would step in to establish a new political system under a new flag or ideology that was welcomed by a substantial portion of the native population. More rarely, the change took place through a revolution, where the ruling elite was overthrown by political dissidents intent on promoting new ideas based on Truth and Justice.

In the modern era, we are witnessing the gradual emergence of a global political system dominated by the world's major nation states, who are all permanent members of the

United Nations Security Council: U.S., Russia, China, Britain, and France. All five of these countries, along with other space faring nations, have been complicit in a global system of secrecy where they, along with major corporations, hide the truth about the ancient history of Antarctica, the space programs first established there by Nazi Germany, and the gross injustices currently occurring there in highly classified programs.

In the early 1990's, Alex Collier said that he had been contacted by extraterrestrials from the Andromeda constellation who said they had arrived to preempt a galactic tyranny that would emerge 350 years in our future. The Andromedans allegedly traced the source of this future galactic tyranny back to Earth in our present time. Collier immediately began publicly discussing the involvement of Draconian Reptilian extraterrestrials in human affairs, the injustices being committed against the human populace, and the secret agreements with national elites that made this all possible:

> Now, in our galaxy there are many councils. I don't know everything about all those councils, but I do know about the Andromedan council, which is a group of beings from 139 different star systems that come together and discuss what is going on in the galaxy. It is not a political body. What they have been recently discussing is the tyranny in our future, 357 years from now, because that affects everybody. Apparently what they have done, through time travel, is that they have been able to figure out where the significant shift in energy occurred that causes the tyranny 357 years in our future. They have traced it back to our solar system, and they have been able to further track it down to Earth, Earth's moon and Mars. Those three places. [583]

If Collier's information is accurate, it's significant that the Earth, Moon and Mars are all identified as sources of a future galactic tryanny. Mars and the Moon in particular are locations where the Fourth Reich's Secret Space Program, along with the transnational "Interplanetary Corporate Program", have established a powerful presence. In particular, Martian industrial colonies allegedly produce advanced technologies for up to 900 different extraterrestrial civilizations.[584] Antarctica's hidden spaceports are the primary location connecting Earth, the Moon and Mars in secret space programs that might constitute a future global/galactic threat.

Based on the information presented in the preceding chapters, I firmly believe that events in Antarctica constitute a form of cancerous growth that can quickly evolve into a global tyranny, which could conceivably evolve further into the galactic tyranny that Collier warned about. Hidden Antarctica bases using slave labor can easily morph into a global phenomenon if left unchecked. Such tyranny would eventually weaken the integrity of the planet's political systems, thereby inviting either an external invasion by extraterrestrials posing as global liberators, or a popular revolution where "full disclosure" becomes the slogan for a new global system based on greater Truth and Justice for all. Therefore, as a matter of national and global security, it is vitally important that the truth about Antarctica be exposed in order to prevent the growth of the unjust practices occurring there, which could potentially morph into a cancerous growth devastating the entire global body politic.

It is also possible that the truth about Antarctica will be revealed as an inevitable byproduct from natural geological processes. As described in chapter 12, Antarctica's buried and submerged volcanoes are heating up west Antarctica, leading to dramatic melting of the ice shelves. Not only will such heating lead to Antarctica's hidden history and corporate run space program being eventually exposed for all the world to see; but may, more disturbingly, lead to a Pole Shift. It would be truly ironic if the exposure of these long-buried secrets proves to be the precise information needed for humanity to escape the cataclysmic effects of a Pole Shift.

Antarctica is a land of enormous size, mystery and challenges. Once the truth about what has transpired there in the remote past and more recently becomes openly known, Antarctica will have the potential to awaken humanity from a deep, deep slumber, caused by the willful suppression of our true history. Full disclosure of Antarctica's history, and current events involving multiple space programs and transnational corporations, will help tremendously in transforming our planet in ways that both startles the imagination and opens up the door for humanity to become a fully liberated member of our greater galactic community.

ACKNOWLEDGMENTS

This book is the result of historical research combined with an analysis of the testimonies of different government/military "insiders" who gained first-hand knowledge about events in Antarctica dating from antiquity to the modern era. The testimonies of these insiders provided the critical element necessary for making sense of otherwise obscure events and documents, which only together cast light on what really has and is transpiring in Antarctica today, and why it is so important to understand.

I am very thankful to the late William Tompkins (1923-2017) who shared much information about Antarctica in our private interviews, which revealed how important Antarctica was in the historic development of secret space programs. His first-hand participation in a U.S. Navy espionage program that gained valuable intelligence of events in Antarctica during World War II, and the Navy's subsequent activities there, were especially helpful in understanding the continent's history. My deep appreciation also goes to another insider, Corey Goode, for the private briefings he gave me which expanded my understanding of recent developments in Antarctica. His personal experiences offered first-hand accounts of what lay buried deep under the ice sheets, making them invaluable. It was Bill and Corey's respective testimonies that opened my eyes to the significance of Antarctica, and the importance of understanding the historical evolution of secret space programs, along with the German and U.S. corporate collaboration that made these developments possible.

I also owe a debt of gratitude to David Wilcock for his *Cosmic Disclosure* interviews with both Corey and Bill, providing invaluable details about ongoing events in Antarctica. In addition, David's interviews with Dr. Pete Peterson about this man's personal experiences in Antarctica were again very helpful in providing further eyewitness corroboration for some of Corey and Bill's claims of what happened there.

Thanks also to veteran UFO researcher, Linda Moulton Howe, for her interviews with "Brian", a former U.S. Navy flight engineer employed in Antarctic for 17 years, who provided more valuable information about events there. I am very grateful to Kathryn "Katie" Leishman, a veteran investigative reporter, for her assistance in tracking down and interviewing a number of individuals for this book who had previously worked in Antarctica, as well as members of a Congressional delegation who traveled there in 2002. Further, she found a former colleague of "Brian" who confirmed that this whistleblower had indeed worked in Antarctica and was a very credible witness.

My warm thanks go to Rene McCann for her generous gift of designing the book's cover which beautifully illustrates the secret agreements that underscore events in Antarctica. Additionally, I am grateful to Gaia TV (Gaia.com) for permission to use graphics that first appeared in *Cosmic Disclosure*. My appreciation also extends to Thomas Keller for permission to use his excellent summary graphic of Corey Goode's claims of multiple secret space programs.

I am thankful to Rene Erik Olsen for permission to use several enhanced photographs taken by George Adamski of his famed 1952 Desert Center encounter with a scout craft and its occupant, Orthon. Thanks likewise to Michel Zirger whose research on the Adamski case first alerted me to Olsen's enhancements.

Much appreciation goes to A. Hughes for copy editing the

final manuscript, and once again assisting in this Secret Space Program book series.

Finally, I am deeply grateful to my wonderful wife and soul mate, Angelika Whitecliff, who once again enthusiastically took on the indispensable role of chief editor for this book. I am truly blessed to have found a life partner who supports my writing and truth telling endeavors in every possible way, and brings her own global transformation projects into my life.

Michael E. Salla, Ph.D.
March 12, 2018

ABOUT THE AUTHOR

Dr. Michael Salla is an internationally recognized scholar in global politics, conflict resolution and U.S. foreign policy. He has taught at universities in the U.S. and Australia, including American University in Washington, DC. Today, he is most popularly known as a pioneer in the development of the field of 'exopolitics'; the study of the main actors, institutions and political processes associated with extraterrestrial life.

He has been a guest speaker on hundreds of radio and TV shows, and featured at national and international conferences. His Amazon best selling *Secret Space Program* book series has made him a leading voice in the Truth Movement, and over 5000 people a day visit his website for his most recent articles.

Dr. Salla's Website: www.exopolitics.org

[1]ENDNOTES

[1] Michael Salla, *Insiders Reveal Secret Space Programs and Extraterrestrial Alliances* (Exopolitics Insitute, 2015)

CHAPTER ONE

[2] See James Pool and Suzanne Pool, *Who Financed Hitler: The Secret Funding of Hitler's Rise to Power 1919-1933* (The Dial Press, 1978) p. 7.
[3] For a summary of these events, see Peter Moon, *The Black Sun: Montauk's Nazi-Tibetan Connection* (Skybooks, 1997) pp. 172ff.
[4] Diodorus Siculus, *Library of History, Book II*, pp. 38-41. Available online at: http://penelope.uchicago.edu/Thayer/E/Roman/Texts/Diodorus_Siculus/2B*.html#note36
[5] See Hyboreans, Wikipedia, https://en.wikipedia.org/wiki/Hyperborea (accessed 6/2/17).
[6] James Pool and Suzanne Pool, *Who Financed Hitler,* p. 7.
[7] "Maria Orsic" http://1stmuse.com/maria_orsitsch/ (accessed 6/6/15).
[8] Wikipedia, "Vril," http://en.wikipedia.org/wiki/Vril (accessed 6/5/15).
[9] Edward Bulwer Lyton, *Vril: the Coming Race*, Chapter 16, available online at: http://www.sacred-texts.com/atl/vril/vrl15.htm (accessed 5/28/17).
[10] Louis Pauwels and Jacques Bergier, *The Morning of the Magicians* (Destiny Books, 2009 [1960]) p. 254.
[11] Cited in Nicholas-Clark, *The Occult Roots of Nazism: Secret Aryan Cults and their Influence on Nazi Ideology* (Tauris Parke Paperbacks, 2004), 219.
[12] Louis Pauwels and Jacques Bergier, *The Morning of the Magicians*, p. 256.
[13] This is how Goodrick-Clark summarizes the perspective of Pauwels and Bergier, a viewpoint he strongly contests as "fallacious". Nicholas-Clark, *The Occult Roots of Nazism*, p. 219.
[14] "Maria Orsic" http://1stmuse.com/maria_orsitsch/ (accessed 6/6/15).
[15] "Maria Orsic" http://1stmuse.com/maria_orsitsch/ (accessed 6/6/15).
[16] James Pool and Suzanne Pool, *Who Financed Hitler,* p. 8.
[17] See Michael Salla, *Insiders Reveal Secret Space Programs and Extraterrestrial Alliances*
[18] "British Intelligence Objectives Sub-Committee: Final Report #1043," http://tinyurl.com/yd2h6ιιlιlι (accessed 8/10/17).
[19] Paul La Violette, *Secrets of Antigravity Propulsion: Tesla, UFOs and Classified Aerospace Technology* (Bear and Co., 2008) p 9.
[20] T.T. Brown, "How I Control Gravitation," *Science & Invention* (August 1929) / *Psychic Observer* 37(1) http://www.rexresearch.com/gravitor/gravitor.htm (accessed on 6/10/15).
[21] James Pool and Suzanne Pool, *Who Financed Hitler,* p. 9.

[22] See James Pool and Suzanne Pool, *Who Financed Hitler,* p. 9.

[23] Jim Marrs, The Rise of the Fourth Reich, p. 19.

[24] James Pool and Suzanne Pool, *Who Financed Hitler,* pp. 9-10.

[25] See James Pool and Suzanne Pool, *Who Financed Hitler,* p. 21.

[26] See James Pool and Suzanne Pool, *Who Financed Hitler,* p. 32.

[27] James Pool and Suzanne Pool, *Who Financed Hitler,* p. 35.

[28] James Pool and Suzanne Pool, *Who Financed Hitler,* p. 8.

[29] See Jan Van Helsing, *Secret Societies and Their Power,* ch. 20, available online at: http://tinyurl.com/ybmezltc

[30] James Pool and Suzanne Pool, *Who Financed Hitler,* p. 11.

[31] James Pool and Suzanne Pool, *Who Financed Hitler,* p. 11.

[32] Nicholas-Clark, *The Occult Roots of Nazism,* p. 221

[33] Cited in Nicholas Goodrick-Clark, *The Occult Roots of Nazism,* p. 219.

[34] Nicholas Goodrick-Clark critiques multiple sources believing in the continuation of the Thule Society in *The Occult Roots of Nazism,* pp. 221-22..

[35] Ian Kershaw, *Hitler: 1889-1936 Hubris,* (W. W. Norton & Company, 2000) pp. 138-139.

[36] Cited by Jan Van Helsing, *Secret Societies and Their Power,* ch. 20, available online at: http://tinyurl.com/ybmezltc Exhaustive lists such as this are strongly disputed by Nicholas Goodrick-Clark, *The Occult Roots of Nazism,* p. 221.

[37] James Pool and Suzanne Pool, *Who Financed Hitler,* pp. 19,21.

[38] James Pool and Suzanne Pool, *Who Financed Hitler,* pp. 32-35.

[39] For discussion of the influence of Haushofer on Hitler, see Jim Marrs, *The Rise of the Fourth Reich,* p. 37.

[40] Jim Marrs, *The Rise of the Fourth Reich,* p. 37.

[41] James Pool and Suzanne Pool, *Who Financed Hitler,* p. 37.

[42] "Peace Treaty of Versailles,": http://net.lib.byu.edu/~rdh7/wwi/versa/versa4.html

[43] James Pool and Suzanne Pool, *Who Financed Hitler,* p. 73.

[44] James Pool and Suzanne Pool, *Who Financed Hitler,* p. 73-74.

[45] Ian Colvin, *Master Spy: The Incredible Story of Admiral Wilhelm Canaris, Who, While Hitler's Chief of Intelligence, Was a Secret Ally of the British* (Uncommon Valor Press, 2014). Kindle Edition, Locations 300-302.

CHAPTER TWO

[46] "Who Financed Adolf Hitler? http://reformation.org/wall-st-ch7.html A different date, 1916, is cited by John Loftus, "How The Bush Family Made Its Fortune From The Nazis: The Dutch Connection," http://www.rense.com/general26/dutch.htm

[47] John Loftus, "How The Bush Family Made Its Fortune From The Nazis: The Dutch Connection," http://www.rense.com/general26/dutch.htm

[48] James Pool, *Who Financed Hitler: The Secret Funding of Hitler,* p. 107.

[49] William Shirer, *The Rise and Fall of the Third Reich: A History of Nazi Germany* (Simon & Schuster, 2011) p. 144.

[50] Fritz Thyssen, cited online at: http://spartacus-educational.com/GERthyssen.htm

[51] Allen Dulles, the Nazis, and the CIA, http://www.panshin.com/trogholm/secret/rightroots/dulles.html

[52] Ben Aris and Duncan Campbell, "How Bush's grandfather helped Hitler's rise to power," https://www.theguardian.com/world/2004/sep/25/usa.secondworldwar

[53] Webster Tarply, *George Bush: The Unauthorized Biography*, chapters available online at: http://tarpley.net/online-books/george-bush-the-unauthorized-biography/chapter-2-the-hitler-project/

[54] Ben Aris and Duncan Campbell, "How Bush's grandfather helped Hitler's rise to power," https://www.theguardian.com/world/2004/sep/25/usa.secondworldwar

[55] Webster Tarply, *George Bush: The Unauthorized Biography*, chapters available online at: http://tarpley.net/online-books/george-bush-the-unauthorized-biography/chapter-2-the-hitler-project/

[56] James Srodes, *Allen Dulles: Master of Spies* (Regnery Publishing Inc., 1999), p. 142.

[57] Signatories of the petition identified by Jim Marrs, *The Rise of the Fourth Reich*, p. 21.

[58] Jim Marrs, *The Rise of the Fourth Reich*, p. 21.

[59] Jim Marrs, *The Rise of the Fourth Reich*, p. 21.

[60] For Allen Dulles meeting with Hitler, see James Srodes, *Allen Dulles: Master of Spies,* p. 163.

[61] Webster Tarpley, *George Bush: The Unauthorized Biography*, ch 2, available online at: http://tarpley.net/online-books/george-bush-the-unauthorized-biography/chapter-2-the-hitler-project/

[62] James Srodes, *Allen Dulles: Master of Spies*, p. 164.

[63] "Henry Ford - The Dearborn Independent and Anti-Semitism," http://www.liquisearch.com/henry_ford/the_dearborn_independent_and_anti-semitism

[64] For discussion of Henry Ford's anti-semitism see Neil Baldwin, *Henry Ford and the Jews: The Mass Production of Hate* (Public Affairs, 2002).

[65] "Henry Ford - The Dearborn Independent and Anti-Semitism," http://www.liquisearch.com/henry_ford/the_dearborn_independent_and_anti-semitism

[66] Jim Marrs, *The Rise of the Fourth Reich*, p. 31.

[67] Jim Marrs, *The Rise of the Fourth Reich*, p. 23.

[68] Jim Marrs, *The Rise of the Fourth Reich*, p. 25.

[69] For detailed description of U.S. Banks investing in Nazi Germany, see Jim Marrs, *The Rise of the Fourth Reich*, pp. 29-30.

[70] Cited by Jim Marrs, *The Rise of the Fourth Reich*, pp. 34-35.

[71] Jim Marrs, *The Rise of the Fourth Reich*, p. 34.

[72] Charles Higham, *Trading with the Enemy: The Nazi-American Money Plot 1933-1949* (Barnes and Noble, 1983). p. xv.

[73] For description of different national versions of "Trading with the Enemy Act," see Wikipedia, https://en.wikipedia.org/wiki/Trading_with_the_Enemy_Act

[74] Charles Higham, *Trading with the Enemy*, xv.

[75] Charles Higham, *Trading with the Enemy*, xv.

[76] Private interview with William Tompkins, July 30, 2017.

[77] Charles Higham, *Trading with the Enemy*, xv.

[78] ITT Corporation, *Wikipedia*, https://en.wikipedia.org/wiki/ITT_Corporation#German_subsidiaries_in_the_Nazi_period (accessed 6/6/17)

[79] Private interview with William Tompkins, July 30, 2017.

[80] Another source describing the divided loyalties of ITT is Anthony Sutton. Chapter five of his book *Wall Street and the Rise of Hitler* (2000) is titled,"I.T.T. Works Both Sides of the War". Available online at: http://www.bibliotecapleyades.net/sociopolitica/wall_street/chapter_05.htm (accessed 11/15/17).

[81] For discussion of the Vril Spacecraft see Michael Salla, *Insiders Reveal Secret Space Programs and Extraterrestrial Alliances*, 65-68.

CHAPTER THREE

[82] See the "Antarctic Enigma," http://www.bibliotecapleyades.net/tierra_hueca/esp_tierra_hueca_6c.htm (accessed 8/7/15).

[83] Interview with William Tompkins, February 25, 2016. Transcript available at: http://exopolitics.org/interview-transcript-reptilian-aliens-helped-germans-establish-space-program-in-antarctica/

[84] Private Interview with William Tompkins, April 30, 2017.

[85] Interview with Corey Goode, May 19, 2014 "Corporate bases on Mars and Nazi infiltration of US Secret Space Program," http://exopolitics.org/corporate-bases-on-mars-and-nazi-infiltration-of-us-secret-space-program/ (accessed 6/30/15).

[86] Chris Pash, "Scientists are closing in on warm caves under Antarctica which could support secret life", https://www.businessinsider.com.au/scientists-are-closing-in-on-warm-caves-under-antarctica-which-could-support-secret-life-2017-9 (accessed 11/8/17).

[87] Robert McKie, "Scientists discover 91 volcanoes below Antarctic ice sheet", https://www.theguardian.com/world/2017/aug/12/scientists-discover-91-volcanos-antarctica

[88] Cited in "Admiral Byrd's 1939 Antartic And ... The Mysterious Snow Cruiser," (accessed 8/7/15). http://www.bibliotecapleyades.net/tierra_hueca/esp_tierra_hueca_18.htm

[89] Private Interview with William Tompkins, April 30, 2017.

[90] I describe Goode's views about German secret societies and their role in secret space program developments in *Insiders Reveal Secret Space Programs and Extraterrestrial Alliances* (Exopolitics Institute, 2015) 75-78.

[91] "Antarctic Enigma," http://www.bibliotecapleyades.net/tierra_hueca/esp_tierra_hueca_6c.htm (accessed 10/29/16).

[92] "The Antarctic Survival Myth," http://www.bibliotecapleyades.net/antarctica/antartica22.htm

[93] "The Antarctic Survival Myth," http://www.bibliotecapleyades.net/antarctica/antartica22.htm

[94] "Ancient Antarctica Was As Warm As Today's California," http://atlanteangardens.blogspot.com/2014/04/ancient-antarctica-was-as-warm-as.html. See also "The world of H.C. Berann", http://www.berann.com/panorama/index.html (accessed 1/16/2018).

[95] "In a Scientific and Engineering Breakthrough, NSF-funded Team Samples Antarctic Lake Beneath the Ice Sheet," https://www.nsf.gov/news/news_images.jsp?org=NSF&cntn_id=126697 (accessed 10/17/17)

[96] "Ancient Antarctica Was As Warm As Today's California," http://atlanteangardens.blogspot.com/2014/04/ancient-antarctica-was-as-warm-as.html

[97] Michael Mueller, *Canaris: The Life and Death of Hitler's Spymaster* (Naval Institute Press, 2007) p. 242.

[98] Michael Mueller describes Canaris new position in, *Canaris: The Life and Death of Hitler's Spymaster*, p. 242.

[99] Michael Mueller describes Canaris new position in, *Canaris: The Life and Death of Hitler's Spymaster*, p. 244.

[100] The details of Canaris' imprisonment, trial and execution are described by Michael Mueller in, *Canaris: The Life and Death of Hitler's Spymaster*, pp. 251-58.

[101] William Tompkins, "Interview Transcript – Reptilian Aliens Helped Germans Establish Space Program in Antarctica," http://exopolitics.org/interview-transcript-reptilian-aliens-helped-germans-establish-space-program-in-antarctica/

[102] Henry Stevens, *Hitler's Flying Saucers: A Guide to German Flying Disks of the Second World War* (Adventures Unlimited Press, 2003)

[103] See Henry Stevens, *Hitler's Flying Saucers*, pp. 55-68.

[104] See Henry Stevens, *Hitler's Flying Saucers*, pp. 144-65.

[105] Document is included in Henry Stevens, *Hitler's Flying Saucers*, p. 151. Available online at: http://p3.pstatp.com/large/363900043f7594da86eb (accessed 11/14/2017)

[106] CIA document available online at: http://alien-ufo-research.com/documents/cia/german-nazi-ufo-flying-saucer-examined-by-cia.php (accessed 11/14/2017)

[107] A brief biography of Vladimir Terziski is available online at: http://www.whale.to/b/terziski_h.html (accessed on 11/14/17)

[108] Vladimir Terziski presented his findings in a 1992 workshop available online at: https://youtu.be/MPBvHjuJtD8

[109] "Rob Arndt, "Haunebu – H-Great, Hanueburg Device 1939-1945)" http://discaircraft.greyfalcon.us/HAUNEBU.htm (accessed on 11/13/17)

[110] "Rob Arndt, http://www.bibliotecapleyades.net/ufo_aleman/esp_ufoaleman_6.htm (accessed 11/13/17)

[111] "Rob Arndt, "Haunebu – H-Great, Hanueburg Device 1939-1945)" http://discaircraft.greyfalcon.us/HAUNEBU.htm (accessed 11/13/17)

[112] Henry Stevens, *Hitler's Suppressed and Still–Secret Weapons, Science and Technology*, pp. 122-30.

[113] Henry Stevens, *Hitler's Suppressed and Still–Secret Weapons, Science and Technology*, pp. 126.

[114] "Rob Arndt, "Haunebu – H-Great, Hanueburg Device 1939-1945)" http://discaircraft.greyfalcon.us/HAUNEBU.htm (accessed 11/13/17)

[115] "Rob Arndt, "Haunebu – H-Great, Hanueburg Device 1939-1945)" http://discaircraft.greyfalcon.us/HAUNEBU.htm (accessed 11/13/17)

[116] "Dornier Flugzeugwerke", Wikipedia, https://en.wikipedia.org/wiki/Dornier_Flugzeugwerke (accessed 11/13/17)

[117] "Rob Arndt, "Haunebu – H-Great, Hanueburg Device 1939-1945)" http://discaircraft.greyfalcon.us/HAUNEBU.htm (accessed 11/13/17)

[118] "Rob Arndt, "Haunebu – H-Great, Hanueburg Device 1939-1945)" http://discaircraft.greyfalcon.us/HAUNEBU.htm (accessed 7/3/17)

[119] Henry Stevens, *Hitler's Suppressed and Still-Secret Weapons, Science and Technology* (Adventures Unlimited Press, 2007) p. 61.

[120] Henry Stevens, *Hitler's Suppressed and Still-Secret Weapons, Science and Technology*, p. 207.

[121] For discussion of I.G. Farben's history in pioneering medical research, see Diarmuid Jeffreys, *Hell's Cartel: IG Farben and the Making of Hitler's War Machine* (Metropolitan Books, 2010)

[122] "Cosmic Disclosure: SSP Testimonials with William Tompkins", https://spherebeingalliance.com/blog/transcript-cosmic-disclosure-ssp-testimonials-with-william-tompkins.html (accessed 11/15/17)

[123] Transcript of Cosmic Disclosure interview, Season 2, Episode 5, http://www.stillnessinthestorm.com/2015/10/cosmic-disclosure-season-2-episode-5.html

[124] For discussion of the Rudolf Hess peace mission to Britain see Jim Marrs, *The Rise of the Fourth Reich*, pp. 36-49.

[125] Jim Marrs, *The Rise of the Fourth Reich*, p. 108.

[126] The full title of the Red House Report is "US Military Intelligence report EW-Pa 128," and it is available online at: https://glwdocuments.wordpress.com/1944/11/27/the-red-house-report-07-nov-1944/

[127] Paul Manning, *Martin Bormann: Nazi in Exile* (Lyle Stuart,1980).

[128] "US Military Intelligence report EW-Pa 128," https://glwdocuments.wordpress.com/1944/11/27/the-red-house-report-07-nov-1944

[129] "US Military Intelligence report EW-Pa 128," https://glwdocuments.wordpress.com/1944/11/27/the-red-house-report-07-nov-1944

[130] Jim Marrs, *The Rise of the Fourth Reich*, p. 111.

[131] Jim Marrs, *The Rise of the Fourth Reich*, p. 113.

[132] War Department, "Elimination of German Resources for War: I.G. Farben Material Submitted by the War Department" (Dec 1945), available online at: http://www.profit-over-life.org/books/books.php?book=30&pageID=13&expand=no&addPage=0 (accessed 11/8/17)

[133] Paul Manning, *Martin Bormann: Nazi in exile,* (Kindle Book Location 1723).

[134] Jim Marrs, *The Rise of the Fourth Reich*, p. 116.

[135] Jim Marrs, *The Rise of the Fourth Reich*, p. 117.

[136] Paul Manning, *Martin Bormann: Nazi in exile* (Kindle Locations 1718-1720).

[137] Paul Manning, *Martin Bormann: Nazi in exile* (Kindle Locations 2065-2067).

[138] See Joseph Farrell, *Nazi International: The Nazis' Postwar Plan to Control Finance, Conflict, Physics and Space* (Adventures Unlimited Press, 2008) Kindle Edition, Location 2672 of 7218.

[139] Paul Manning, *Martin Bormann: Nazi in Exile* (CreateSpace, 1981)

[140] Harry Cooper, *Hitler in Argentina: The Documented Truth of Hitler's Escape from Berlin* (CreateSpace, 2014).

[141] Document available at CIA website, https://www.cia.gov/library/readingroom/docs/HITLER%2C%20ADOLF_0003.pdf (accessed 11/09/17).

CHAPTER FOUR

[142] For extensive discussion of such an exodus to Antarctica and South America, see Joscelyn Godwin, *Arktos: The Polar Myth in Science, Symbolism, and Nazi Survival;* Jim Marrs, Alien Agenda, 107-13; and also see Branton, *The Omega Files; Secret Nazi UFO Bases Revealed* (Inner Light Publications, 2000). Available online at: http://www.think-about-it.com (accessed 6/30/15).

[143] Private interview with William Tompkins, April 30, 2017.

[144] Interview transcript, "Cosmic Disclosure: SSP Testimonials with William Tompkins," http://spherebeingalliance.com/blog/transcript-cosmic-disclosure-ssp-testimonials-with-william-tompkins.html

[145] Richard K. Wilson and Sylvan Burns, *Secret Treaty: The United States Government and Extra-terrestrial Entities* (N.A.R, 1989), cited from: http://www.thewatcherfiles.com/alien-treaty.htm (accessed 6/30/15).

[146] Interviewed by Linda Moulton Howe, Earthfiles, http://www.earthfiles.com/news.php?ID=1464&category=Real%20X-Files (accessed 4/4/15).

[147] Interviewed by Linda Moulton Howe, Earthfiles, http://www.earthfiles.com/news.php?ID=1464&category=Real%20X-Files (accessed 4/4/15).

[148] For extensive discussion of the advanced technology left by Nazi Germany, see "Secrets of the Third Reich," https://www.bibliotecapleyades.net/ufo_aleman/esp_ufoaleman_8a.htm (accessed 11/18/2017).

[149] For discussion of how senior Nazi began transferring funds and resources through South America, see Marrs, *Alien Agenda,* 107-113.

[150] "Antarctic Enigma," http://www.bibliotecapleyades.net/tierra_hueca/esp_tierra_hueca_6c.htm (accessed 8/7/15).

[151] Private Interview with William Tompkins, April 30, 2017.

[152] In introduction to Renato Vesco and David Hatcher Childress, *Man-Made UFOs 1944-1994: 50 Years of Suppression* (AUP Publishers, 1994/2005).

[153] Email Interview with Corey Goode, May 20, 2015, http://exopolitics.org/corporate-bases-on-mars-and-nazi-infiltration-of-us-secret-space-program/

[154] Transcript of Interview with William Tompkins, April 14, 2016, http://exopolitics.org/interview-transcript-reptilian-aliens-helped-germans-establish-space-program-in-antarctica/

[155] For discussion of Nazi developed Saucers being witnessed during the post-World War period, see "Secrets of the Third Reich," https://www.bibliotecapleyades.net/ufo_aleman/esp_ufoaleman_8a.htm (accessed 11/18/2017).

[156] "The Antarctic Enigma," http://www.bibliotecapleyades.net/tierra_hueca/esp_tierra_hueca_6c.htm (accessed 8/7/15). For further references to Operation Highjump see, Branton, *The Omega Files.* Available online at http://www.think-aboutit.com/Omega/files/omega3.htm (accessed 7/6/15).

[157] "The Antarctic Enigma," http://www.bibliotecapleyades.net/tierra_hueca/esp_tierra_hueca_6c.htm (accessed 8/7/15).

[158] Private Interview with William Tompkins, April 30, 2017.

[159] Presentation by Donald Ware on "Sharing Earth: Various Intelligent Species", International UFO Congress, https://youtu.be/mL_pTaiJZHo?t=1h8m50s (accessed 11/15/17).

[160] John Livermore, "Goering's Hi Tech Mission: The German Antarctic Expedition 1938-9," http://johnlivermore.com/files/GERMAN%20ANTARCTIC%20EXPEDITION%201938.doc (accessed 8/7/15).

[161] Cited in "Admiral Byrd's 1939 Antartic And ... The Mysterious Snow Cruiser,"
http://www.bibliotecapleyades.net/tierra_hueca/esp_tierra_hueca_18.htm (accessed 8/7/15).

[162] See "The Antarctic Enigma,"
http://www.bibliotecapleyades.net/tierra_hueca/esp_tierra_hueca_6c.htm (accessed 8/7/15).

[163] Lee Van Atta's article is available online at:
https://www.bibliotecapleyades.net/imagenes_antartica/antartica22_03.gif (accessed 11/13/17).

[164] Translation supplied via Linda Moulton Howe and Duncan Roads, "Operation Highjump Photos from Anonymous Source"
https://www.earthfiles.com/news.php?ID=2577&category=Science, (accessed 11/13/17).

[165] Translation supplied via Linda Moulton Howe and Duncan Roads, "Operation Highjump Photos from Anonymous Source"
https://www.earthfiles.com/news.php?ID=2577&category=Science, (accessed 11/13/17).

[166] "Third Reich - Operation UFO (Nazi Base In Antarctica) Complete Documentary" https://youtu.be/MwUpPwyyvLw (accessed 7/6/15).

[167] Our Real "War of the Worlds," http://www.newdawnmagazine.com/special-issues/new-dawn-special-issue-vol-6-no-5 (accessed 7/6/15).

[168] Our Real "War of the Worlds," http://www.newdawnmagazine.com/special-issues/new-dawn-special-issue-vol-6-no-5 (accessed 7/6/15).

[169] Wikipedia, "List of torpedo boats of the United States Navy"
https://en.wikipedia.org/wiki/List_of_torpedo_boats_of_the_United_States_Navy (accessed 8/7/15).

[170] Wikipedia, "USS Maddox," https://en.wikipedia.org/wiki/USS_Maddox

[171] Our Real "War of the Worlds," http://www.newdawnmagazine.com/special-issues/new-dawn-special-issue-vol-6-no-5 (accessed 7/6/15).

[172] See Raymond W. Bernard, *The hollow Earth : the greatest geographical discovery in history made by Admiral Richard E. Byrd in the mysterious land beyond the Poles--the true origin of the flying saucers* (Bell Publishing Co.). For online discussion of ET sightings in Antarctica region, see "Antarctic Enigma,"
http://www.bibliotecapleyades.net/tierra_hueca/esp_tierra_hueca_6c.htm (accessed 8/7/15).

[173] Stein was inteviewed by Linda Moulton Howe, Earthfiles,
http://www.earthfiles.com/news.php?ID=1464&category=Real%20X-Files (accessed 9/24/14).

[174] Quoted in an interview of Admiral Byrd by Lee van Atta, "On Board the Mount Olympus on the High Seas" *El Mercurio*, (Santiago, Chile, March 5, 1947). See "The Antarctic Enigma,"
http://www.bibliotecapleyades.net/tierra_hueca/esp_tierra_hueca_6c.htm (accessed 8/7/15).

[175] Interview with Corey Goode, May 19, 2014 "Corporate bases on Mars and Nazi infiltration of US Secret Space Program," http://exopolitics.org/corporate-bases-on-mars-and-nazi-infiltration-of-us-secret-space-program/ (accessed 6/30/15)

[176] Admiral Richard Byrd's Diary, available online at: http://www.bibliotecapleyades.net/tierra_hueca/esp_tierra_hueca_2d.htm

[177] William Tompkins, *Selected by Extraterrestrials*, p. 139.

[178] For discussion of converted nuclear submarines, see Michael Salla, *The US Navy's Secret Space Program and Nordic Extraterrestrial Alliance* (Exopolitics Consultants, 2017) 148-49.

[179] Rense Radio Interview with William Tompkins with Maj. George Filer & Frank Chille – May 4, 2016. Transcript available at: http://spherebeingalliance.com/blog/the-amazing-story-continues-part1.html

[180] Interview with Corey Goode, May 19, 2015 "Corporate bases on Mars and Nazi infiltration of US Secret Space Program," http://exopolitics.org/corporate-bases-on-mars-and-nazi-infiltration-of-us-secret-space-program/ (accessed 6/30/15).

[181] Jim Marrs, *The Rise of the Fourth Reich*, p. 151-52.

[182] Jim Marrs, *The Rise of the Fourth Reich*, p. 153.

[183] Private Interview with William Tompkins, April 30, 2017.

[184] Jim Marrs, *The Rise of the Fourth Reich*, pp. 154-55.

[185] Private Interview with William Tompkins, April 30, 2017.

[186] Interview with Corey Goode, May 19, 2014 "Corporate bases on Mars and Nazi infiltration of US Secret Space Program," http://exopolitics.org/corporate-bases-on-mars-and-nazi-infiltration-of-us-secret-space-program/ (accessed 6/30/15). http://www.stargate-chronicles.com/site/

[187] Clark McClelland, *The Stargate Chronicles*, chapter 28, http://www.stargate-chronicles.com/site/

[188] Clark McClelland, *The Stargate Chronicles*, chapter 32, http://www.stargate-chronicles.com/site/

[189] Interview with Corey Goode, May 19, 2014 "Corporate bases on Mars and Nazi infiltration of US Secret Space Program," http://exopolitics.org/corporate-bases-on-mars-and-nazi-infiltration-of-us-secret-space-program/ (accessed 6/30/15).

CHAPTER FIVE

[190] Cited by Art Campbell, http://www.ufocrashbook.com/eisenhower.html

[191] For discussion of the 1954 Edwards Air Force Base meeting, see Michael Salla, *Galactic Diplomacy: Getting to Yes with ET* (Exopolitics Institute, 2013].

[192] See Art Campbell, http://www.ufocrashbook.com/eisenhower.html

[193] Bill Kirklin is the author of "Ike and UFOs," that was published anonymously in the *Exopolitics Journal* 2:1 (2007): http://exopoliticsjournal.com/vol-2/vol-2-1-Exp-Ike.htm Kirklin's says he was

told about Ike's visit in late February, this is likely a minor mistake on his part since the visit occurred on February 11, 1955.

[194] Kirklin,"Ike and UFOs:" http://exopoliticsjournal.com/vol-2/vol-2-1-Exp-Ike.htm

[195] Kirklin,"Ike and UFOs:" http://exopoliticsjournal.com/vol-2/vol-2-1-Exp-Ike.htm

[196] Transcript of electrician's family letter – recorded by Art Campbell and played on Jerry Pippin Show - 6/23/08. Available online at: http://www.ipodshows.net/Archives_3rd_quarter_2008.htm

[197] Transcript of Staff Sgt Wykoff interviewed by Art Campbell and played on Jerry Pippin Show - 6/23/08. Available online at: http://www.ipodshows.net/Archives_3rd_quarter_2008.htm

[198] Clark McClelland, *The Stargate Chronicles*, ch. 32, http://www.stargate-chronicles.com/site/ (accessed 10/17/17).

[199] Clark McClelland, *The Stargate Chronicles*, ch. 32, http://www.stargate-chronicles.com/site/ (accessed 10/17/17).

[200] Clark McClelland, *The Stargate Chronicles*, ch. 32, http://www.stargate-chronicles.com/site/ (accessed 10/17/17).

[201] Private Interview with William Tompkins, April 30, 2017.

[202] Interview with Corey Goode, May 19, 2014, "Corporate bases on Mars and Nazi infiltration of US Secret Space Program," http://exopolitics.org/corporate-bases-on-mars-and-nazi-infiltration-of-us-secret-space-program/ (accessed 6/30/15).

[203] Interview with Corey Goode, May 19, 2015 "Corporate bases on Mars and Nazi infiltration of US Secret Space Program," http://exopolitics.org/corporate-bases-on-mars-and-nazi-infiltration-of-us-secret-space-program/ (accessed 6/30/15).

[204] Private Interview with William Tompkins, April 30, 2017.

[205] Clark McClelland, *The Stargate Chronicles*, ch. 15, http://tinyurl.com/ox66j9y (accessed 6/30/15).

[206] Jim Marrs, *The Rise of the Fourth Reich*, p. 156.

[207] Jim Marrs, *The Rise of the Fourth Reich*, p. 156.

[208] Jim Marrs, *The Rise of the Fourth Reich*, p. 158.

[209] Charles Higham, *Trading with the Enemy: The Nazi-American Money Plot 1933-1949*, pp. xiii-xxi.

[210] For detailed discussion of the Dark Fleet, see Michael Salla, *Insiders Reveal Secret Space Programs and Extraterrestrial Alliances*, pp. 117-46.

[211] For further discussion of US Air Force approach to developing a secret space program, see Michael Salla, *The US Navy's Secret Space Program and Nordic Extraterrestrial Alliance*, pp. 205-36.

[212] For further discussion of US Air Force approach to developing a secret space program, see Michael Salla, *The US Navy's Secret Space Program and Nordic Extraterrestrial Alliance*, pp. 205-36.

[213] For discussion of the Los Angeles Air Raid, see chapter one, Michael Salla, *The US Navy's Secret Space Program and Nordic Extraterrestrial Alliance*.

[214] "Twining's 'White Hot' Report," *The Majestic Documents* (Wood and Wood Enterprises, 1998) p. 75.

[215] See Majestic Documents website, http://tinyurl.com/jt49ov3 (accessed 11/9/17)

[216] *The Antarctic Sun*, "IGY +40: From Montparnsasse to McMurdo" , https://antarcticsun.usap.gov/pastIssues/1997-1998/1997_11_01.pdf (accessed 11/9/17).

[217] *The Antarctic Sun*, "IGY +40: From Montparnsasse to McMurdo" , https://antarcticsun.usap.gov/pastIssues/1997-1998/1997_11_01.pdf (accessed 11/9/17).

[218] *The Antarctic Sun*, "Antarctica Prepares for Science," https://antarcticsun.usap.gov/pastIssues/1997-1998/1997_11_29.pdf (accessed 11/9/17).

CHAPTER SIX

[219] For a comprehensive list see: "Holocaust Restitution: German Firms that Used Slave Labor During Nazi Era," http://www.jewishvirtuallibrary.org/german-firms-that-used-slave-labor-during-nazi-era

[220] Albert Speer, *Infiltration: How Heinrich Himmler Schemed to Build an SS Industrial Empire* (McMillan Publishing, 1981) p. 83.

[221] Fritz Sauckel (1894-1946) http://www.erfurt-web.de/Fritz_Sauckel_English (accessed 8/2/17).

[222] Fritz Sauckel (1894-1946) http://www.erfurt-web.de/Fritz_Sauckel_English (accessed 8/2/17).

[223] Fritz Sauckel (1894-1946) http://www.erfurt-web.de/Fritz_Sauckel_English (accessed 8/2/17).

[224] "Financial compensation for Nazi slave laborers, http://www.religioustolerance.org/fin_nazi.htm (accessed 8/2/17).

[225] Albert Speer, *Infiltration*, p. 301.

[226] Albert Speer, *Infiltration*, p. 205.

[227] "Peenemunde – 1943", http://www.globalsecurity.org/wmd/ops/peenemunde.htm (accessed 8/2/17).

[228] Albert Speer, *Infiltration*, p. 205.

[229] Albert Speer, *Infiltration*, p. 209.

[230] Albert Speer, *Infiltration*, pp. 210-11.

[231] Albert Speer, *Infiltration*, p. 219.

[232] Albert Speer, *Infiltration*, p. 227.

[233] Albert Speer, *Infiltration*, p. 218.

[234] Albert Speer, *Infiltration*, p. 219.

[235] Albert Speer, *Infiltration*, pp. 301-302.

[236] Private Interview with William Tompkins, April 17, 2017

[237] For detailed discussion of the different spacecraft developed by Nazi Germany, see Henry Stevens, *Hitler's Flying Saucers: A Guide to German Flying Discs of the Second World War*, 2nd edition (Adventures Unlimited Press, [2003] 2013).

[238] For detailed discussion of the different technological breakthroughs achieved in Nazi Germany, see Henry Stevens, *Hitler's Suppressed and Still-Secret Weapons, Science and Technology* (Adventures Unlimited Press, 2007).

[239] Quoted in an interview of Admiral Byrd by Lee van Atta, "On Board the Mount Olympus on the High Seas" *El Mercurio*, (Santiago, Chile, March 5, 1947). See "The Antarctic Enigma," http://www.bibliotecapleyades.net/tierra_hueca/esp_tierra_hueca_6c.htm (accessed 11/8/17).

[240] See Rudolf Lusar, German Secret Weapons of the Second World War, 2nd edition (N. Spearman. [1956] 1960); and Renato Vesco, *Intercept UFO* (Grove Press, [1968] 1971).

[241] See Henry Stevens, *Hitler's Flying Saucers: A Guide to German Flying Discs of the Second World War*, and Joseph Farrell, *Nazi International: The Nazi Postwar Plan to Control Finance, Conflict, Physics and Space* (Adventures Unlimited Press, 2013).

[242] Henry Stevens, *Hitler's Flying Saucers: A Guide to German Flying Discs of the Second World War* (Kindle Edition, Location 69-72.

[243] Joseph Farrell, *Nazi International,* Kindle Edition, Location 3018 of 7218.

[244] W. A. Harbinson, Introduction to Renato Vesco and David Childress Hatcher, *Man-Made UFOs 1944-1994: 50 Years of Suppression* (Adventures Unlimited Press, 1995) n.p.

[245] William Tompkins interviewed on *Cosmic Disclosure*, February 21 2017. https://spherebeingalliance.com/blog/transcript-cosmic-disclosure-deeper-disclosures-from-william-tompkins.html (accessed 11/9/17).

[246] William Tompkins interviewed on *Cosmic Disclosure*, February 21 2017. https://spherebeingalliance.com/blog/transcript-cosmic-disclosure-deeper-disclosures-from-william-tompkins.html (accessed 11/9/17).

[247] See Michael Mueller, Canaris: The Life and Death of Hitler's Spymaster, (Chatham Publishing, 2007) p. 136; and Kathy Warnes, "Fritz Thyssen Helped Finance the Nazi Party, but Later Changed His Mind", http://windowstoworldhistory.weebly.com/fritz-thyssen-helped-finance-the-nazi-party-but-later-changed-his-mind.html (accessed 11/9/17).

CHAPTER SEVEN

[248] Michel Zirger, *We Are Here: Visitors without a Passport* (Amazon Digital Services, 2017). (Kindle location 5203-10 of 5403).

[249] See "Giuseppe Belluzzo, the Italian engineer" http://www.naziufos.com/2016/03/06/giuseppe-belluzzo-italian-engineer-ufo/

and "Major Lusar, the Saucer Builders, and the test flight",
http://naziufomythos.greyfalcon.us/majorlusar.html (accessed 12/19/17).

[250] See Michael Salla, "US Navy Spies Learned of Nazi Alliance with Reptilian Extraterrestrials during WWII" http://exopolitics.org/us-navy-spies-learned-of-nazi-alliance-with-reptilian-extraterrestrials-during-wwii/ (accessed 12/19/17)

[251] Interview Transcript, "Cosmic Disclosure (S5E10): SSP Testimonials with William Tompkins," http://spherebeingalliance.com/blog/transcript-cosmic-disclosure-ssp-testimonials-with-william-tompkins.html

[252] Interview with Corey Goode, "Corporate bases on Mars and Nazi infiltration of US Secret Space Program", http://exopolitics.org/corporate-bases-on-mars-and-nazi-infiltration-of-us-secret-space-program/ (accessed 12/19/17).

[253] See Michel Zirger *We Are Here* (Kindle Location 903 of 5403).

[254] Michel Zirger *We Are Here* (Kindle Location 526-531 of 5403).

[255] "UFOs The Contacts" https://www.youtube.com/watch?v=kPvvz7O3CKk (accessed 12/19/17).

[256] See Michael Salla, "Did President Kennedy Meet Extraterrestrials" https://www.bibliotecapleyades.net/ciencia/ciencia_flyingobjects71.htm (accessed 12/19/17).

[257] "7 Jan 1956 - Kaimuki, Hawaii, USA", http://thecid.com/ufo/uf07/uf4/074165.htm (accessed 12/19/17).

[258] Willard Layton Wannall, "Wheels Within Wheels," https://library.abundanthope.org/index_htm_files/Wheels%20within%20Wheels-AH.pdf (accessed 12/19/17).

[259] Hawaii UFO Magazine #2 (Limited distribution in Maui, Hawaii).

[260] Reinhold O. Schmidt, *Edge of Tomorrow* (Inner Light, 1975) available online at http://galactic.to/rune/reinhold.html (accessed 12/19/17).

[261] Reinhold O. Schmidt, *Edge of Tomorrow* (Inner Light, 1975) available online at http://galactic.to/rune/reinhold.html (accessed 12/19/17).

[262] Wendelle Stevens, *Message From the Pleiades: The Contact Notes of Eduard Billy Meier*, Volume 1 (UFO Photo Archives, 1988).

[263] Justin Deschamps, "David Wilcock and Corey Goode: History of the Solar System and Secret Space Program - Notes from Consciousness Life Expo 2016 ," http://sitsshow.blogspot.com.au/2016/02/david-wilcock-and-corey-goode-history.html

CHAPTER EIGHT

[264] "Eisenhower's Farewell Speech", available online at: http://mcadams.posc.mu.edu/ike.htm

[265] *Prelude to Leadership: The European Diary of John F. Kennedy: Summer 1945* (Regnery Publishing, 1997).

[266] For discussion on what Forrestal told Kennedy, see Michael Salla, *Kennedy's Last Stand: UFOs, MJ-12, & JFK's Assassination* (Exopolitics Institute, 2013) pp. 11-32.

[267] Charles Higham, *Trading with the Enemy*, 181.

[268] The ties between the Kennedys and the Italian Mafia is documented in Seymour Hersh, *The Dark Side of Camelot* (Back Bay Books, 1998).

[269] Available online at:
http://www.majesticdocuments.com/pdf/kennedy_ciadirector.pdf

[270] For discussion documents supporting the authenticity of the June Memorandum to Dulles, see Michael Salla, *Kennedy's Last Stand: UFOs, MJ-12, & JFK's Assassination*, pp. 107-26.

[271] "Operations Review: The MJ-12 Project," available online at:
http://www.majesticdocuments.com/pdf/mj12opsreview-dulles-61.pdf

[272] "Operations Review: The MJ-12 Project," available online at:
http://www.majesticdocuments.com/pdf/mj12opsreview-dulles-61.pdf

[273] "Operations Review: The MJ-12 Project," available online at:
http://www.majesticdocuments.com/pdf/mj12opsreview-dulles-61.pdf

[274] "Operations Review: The MJ-12 Project," available online at:
http://www.majesticdocuments.com/pdf/mj12opsreview-dulles-61.pdf

[275] Cord Meyer, *Facing Reality: From World Federalism to the CIA* (Harper and Row, 1980) 205-08

[276] Mark Riebling, Wedge: *The Secret War between the FBI and the CIA* (Alfred Knopf, 1994) 327.

[277] Available online at:
http://www.majesticdocuments.com/pdf/burnedmemocoverletter.pdf

[278] Letter addressed to Timothy Cooper, June 23, 1999. Available online at:
http://majesticdocuments.com/pdf/burnedmemocoverletter.pdf

[279] See prefatory comments for burned memo at:
http://www.majesticdocuments.com/documents/1960-1969.php

[280] The burned memorandum is available online at:
http://majesticdocuments.com/documents/1960-1969.php#burnedmemo

[281] See page one of burned memorandum available at:
http://www.majesticdocuments.com/pdf/burnedmemo-s1-pgs1-2.pdf

[282] "John F. Kennedy to Director, CIA,"
http://www.majesticdocuments.com/pdf/kennedy_ciadirector.pdf

[283] See page one of burned memorandum available at:
http://www.majesticdocuments.com/pdf/burnedmemo-s1-pgs1-2.pdf

[284] Cited online at: http://www.majesticdocuments.com/pdf/burnedmemo-s1-pgs3-9.pdf

[285] Available online at: http://www.scribd.com/doc/6404101/JFK-MJ12

[286] John F Kennedy Inaugural Address, January 20, 1961. Source:
http://www.jfklibrary.org/Asset-Viewer/BqXIEM9F4024ntFl7SVAjA.aspx

[287] John F Kennedy Inaugural Address, January 20, 1961. Source:
http://www.jfklibrary.org/Asset-Viewer/BqXIEM9F4024ntFl7SVAjA.aspx

[288] John F Kennedy Inaugural Address, January 20, 1961. Source:
http://www.jfklibrary.org/Asset-Viewer/BqXIEM9F4024ntFl7SVAjA.aspx

[289] Address Before the 18th General Assembly of the United Nations (September 20, 1963). Available at:

http://www.jfklibrary.org/Historical+Resources/Archives/Reference+Desk/Spee ches/JFK/003POF03_18thGeneralAssembly09201963.htm

[290] See: http://history.nasa.gov/moondec.html

[291] Source: http://www.pbs.org/redfiles/moon/deep/interv/m_int_sergei_khrushchev.htm

[292] Available online at: http://www.spacewar.com/news/russia-97h.html

[293] Available online at: http://www.pbs.org/redfiles/moon/deep/moon_deep_inter_frm.htm

[294] Frank Sietzen, "Soviets Planned to Accept JFK's Joint Lunar Mission Offer," http://www.spacewar.com/news/russia-97h.html

[295] Hoagland and Bara, *Dark Mission,* 101.

[296] Available online at: http://tinyurl.com/mejpm4

[297] Available online at: http://tinyurl.com/mejpm4

[298] Available online at: http://www.majesticdocuments.com/pdf/kennedy_cia.pdf

[299] See: http://www.cufon.org/cufon/janp1462.htm

[300] "The Bolender Memo, Oct 20, 1969," http://www.nicap.org/Bolender_Memo.htm

[301] For information on Project Blue Book, go to: http://www.ufocasebook.com/bluebook.html

[302] For rating system used by the founders of the Majestic Documents website, go to: http://majesticdocuments.com/documents/authenticity.php

[303] Available online at: http://tinyurl.com/mejpm4

[304] Leading UFO researcher Allen Hynek claimed that after the departure of Captain Ruppelt, Hynek "Project Blue Book was little more than a public relations exercise." http://en.wikipedia.org/wiki/J._Allen_Hynek

[305] Available online at: http://www.majesticdocuments.com/pdf/kennedy_cia.pdf

[306] NSAM 271 available online at: http://tinyurl.com/mejpm4

[307] Transcript of a Recording of a Meeting Between the President and H. R. Haldeman, the Oval Office, June 23, 1972.

[308] H.R. Haldeman, *The Ends of Power* (Times Books, 1978) p. 39.

[309] "CIA Agent Confesses On Deathbed "I Was Part Of An Assassination Team Of Killing John F. Kennedy"" http://nativestuff.us/2017/08/cia-agent-confesses-on-deathbed-i-was-part-of-an-assassination-team-of-killing-john-f-kennedy-3/ (accessed 11/27/16).

[310] It was also included in Saint John Hunt, *Bond of Secrecy: My Life with CIA Spy and Watergate Conspirator E. Howard Hunt* (Trine Day, 2012).

[311] Linda Moulton Howe, "Part 2: Attorney Douglas Caddy's Assassination Secrets", https://www.earthfiles.com/news.php?ID=2580&category=Environment (accessed 11/27/17).

[312] Siemens website, https://www.siemens.com/global/en/home/company/about.html (accessed 8/12/17).

[313] "Hermann von Siemens," https://everipedia.org/wiki/Hermann_von_Siemens/#ixzz4pZHANpNS (accessed 8/12/17).

[314] Associated Press, "Siemens Offers $12 Million to WWII Slave Labor Victims," http://articles.latimes.com/1998/sep/24/news/mn-26067 (accessed 8/12/17).

[315] Associated Press, "Siemens Offers $12 Million to WWII Slave Labor Victims," http://articles.latimes.com/1998/sep/24/news/mn-26067 (accessed 8/12/17).

[316] Video available online at: https://www.youtube.com/watch?v=9OiZRr9V7Z4 (accessed 8/12/17).

[317] Video available online at: https://www.youtube.com/watch?v=9OiZRr9V7Z4 (timestamp: 11:26-12:00 - accessed 8/12/17).

[318] Video available online at: https://www.youtube.com/watch?v=9OiZRr9V7Z4 (timestamp: 14:00-14:35 - accessed 8/12/17).

[319] Video available online at: https://www.youtube.com/watch?v=9OiZRr9V7Z4 (timestamp:21:47-21:53 - accessed 8/12/17).

[320] Video available online at: https://www.youtube.com/watch?v=9OiZRr9V7Z4(timestamp:22:00-22:23 - accessed 8/12/17)

[321] Video available online at: https://www.youtube.com/watch?v=9OiZRr9V7Z4(timestamp:23:45-24:53 - accessed 8/12/17).

[322] Michael Salla, "Corporate bases on Mars and Nazi infiltration of US Secret Space Program", http://exopolitics.org/corporate-bases-on-mars-and-nazi-infiltration-of-us-secret-space-program/ (accessed 11/16/17).

[323] "William Pawelec's widow reveals national security secrets," http://tinyurl.com/b94kgj4 (accessed 8/12/17).

[324] William Tompkins Interview on Cosmic Disclosure (September 6, 2017) http://spherebeingalliance.com/blog/transcript-cosmic-disclosure-remembering-william-tompkins-disrupting-draco-domination.html (accessed 10/18/17).

[325] William Tompkins Interview on Cosmic Disclosure (September 6, 2017) http://spherebeingalliance.com/blog/transcript-cosmic-disclosure-remembering-william-tompkins-disrupting-draco-domination.html (accessed 10/18/17).

[326] "Questions for Corey Goode on SSP Conflicts and Human Slave Trade," http://exopolitics.org/galactic-human-slave-trade-ai-threat-to-end-with-full-disclosure-of-et-life/ (accessed 10/18/17).

[327] Associated Press, "Siemens Offers $12 Million to WWII Slave Labor Victims," http://articles.latimes.com/1998/sep/24/news/mn-26067 (accessed 8/12/17).

CHAPTER TEN

[328] US State Department, "Antarctic Treaty",
https://www.state.gov/t/avc/trty/193967.htm (accessed 8/13/17).
[329] US State Department, "Antarctic Treaty",
https://www.state.gov/t/avc/trty/193967.htm (accessed 8/13/17).
[330] US State Department, "Antarctic Treaty",
https://www.state.gov/t/avc/trty/193967.htm (accessed 8/13/17).
[331] US State Department, "Antarctic Treaty",
https://www.state.gov/t/avc/trty/193967.htm (accessed 8/13/17).
[332] United States Antarctic Program, Participant Guide 2016-2018 Edition, p. 3.
https://www.usap.gov/USAPgov/travelAndDeployment/documents/ParticipantG
uide_2016-18.pdf (accessed 10/18/17).
[333] United States Antarctic Program, Participant Guide 2016-2018 Edition, p. 2.
https://www.usap.gov/USAPgov/travelAndDeployment/documents/ParticipantG
uide_2016-18.pdf (accessed 10/18/17).
[334] Cited in the Antarctic Sun, https://antarcticsun.usap.gov/pastIssues/1997-
1998/1998_02_07.pdf
[335] US State Department, "The Antarctic Treaty",
https://www.state.gov/documents/organization/81421.pdf (accessed 8/13/17).
[336] Paul Manning, *Martin Bormann: Nazi in Exile* (CreateSpace, 1981)
[337] Harry Cooper, *Hitler in Argentina: The Documented Truth of Hitler's Escape
from Berlin* (CreateSpace, 2014).
[338] *The Antarctic Sun* (2/7/1998) https://antarcticsun.usap.gov/pastIssues/1997-
1998/1998_02_07.pdf (accessed 10/18/17).
[339] For detailed discussion of the Dark Fleet and its origin, see Michael Salla,
Insiders Reveal Secret Space Programs and Extraterrestrial Alliances
(Exopolitics Institute, 2015) 120-46.
[340] See Michael Salla, *The U.S. Navy's Secret Space Program & Nordic
Extraterrestrial Alliance* (Exopolitics Consultants, 2017) pp.181-204.
[341] Email interview with Corey Goode, May 14, 2015,
http://exopolitics.org/corporate-bases-on-mars-and-nazi-infiltration-of-us-secret-
space-program/ (accessed 11/9/17).
[342] "The Erebus Flight Path Controversy",
http://www.erebus.co.nz/Background/TheFlightPathControversy.aspx
[343] "Air New Zealand DC-10 crash into Mt. Erebus",
http://www.southpolestation.com/trivia/history/te901.html (accessed 8/13/17)
[344] "Erebus Disaster," https://nzhistory.govt.nz/culture/erebus-disaster/inquiry
(accessed 8/13/17).
[345] "The Legal Process,"
http://www.erebus.co.nz/Investigation/Legalprocess.aspx (accessed 8/13/17).
[346] "The Legal Process,"
http://www.erebus.co.nz/Investigation/Legalprocess.aspx (accessed 8/13/17).

[347] For description of QANTAS Airways flights to Antarctica, see: https://antarcticsun.usap.gov/pastIssues/1997-1998/1998_02_07.pdf (accessed 8/13/17).

[348] His most recent interview was in November 2017, "Linda Moulton Howe Interview of Naval Officer – Antarctica", https://youtu.be/ZlOPsidcBfo (accessed 11/17/17).

[349] "Navy Engineer Interviewed: I Saw Antarctic UFOs, Aliens and Top-Secret Bases,"
http://www.unsilentmajoritynews.com/navy-engineer-interviewed-i-saw-antarctic-ufos-aliens-and-top-secret-bases-audio/

[350] Brian's responses to my questions were received on October 25, 2017.

[351] Mike Wehner, "Something scorching hot is melting Antarctica from below, and NASA thinks they know what it is,
https://www.yahoo.com/news/something-scorching-hot-melting-antarctica-below-nasa-thinks-180655776.html

[352] "Linda Moulton Howe Interview of Naval Officer – Antarctica", https://youtu.be/ZlOPsidcBfo (accessed 11/17/17).

[353] "Navy Engineer Interviewed: I Saw Antarctic UFOs, Aliens and Top-Secret Bases,"
http://www.unsilentmajoritynews.com/navy-engineer-interviewed-i-saw-antarctic-ufos-aliens-and-top-secret-bases-audio/

[354] Brian's responses to my questions were received on October 25, 2017.

[355] "Navy Engineer Interviewed: I Saw Antarctic UFOs, Aliens and Top-Secret Bases,"
http://www.unsilentmajoritynews.com/navy-engineer-interviewed-i-saw-antarctic-ufos-aliens-and-top-secret-bases-audio/

[356] "Navy Engineer Interviewed: I Saw Antarctic UFOs, Aliens and Top-Secret Bases,"
http://www.unsilentmajoritynews.com/navy-engineer-interviewed-i-saw-antarctic-ufos-aliens-and-top-secret-bases-audio/

[357] "Linda Moulton Howe Interview of Naval Officer – Antarctica", https://youtu.be/ZlOPsidcBfo (accessed 11/17/17).

[358] "Antarctica Retired Navy Flight Engineer Warned by NSA to Stop Talking About Missing Scientists"
https://www.earthfiles.com/news.php?ID=2440&category=Science (accessed 10/19/17).

CHAPTER ELEVEN

[359] Wikipedia, "Lake Vostok," https://en.wikipedia.org/wiki/Lake_Vostok (accessed 8/15/17).

[360] Antarctic Sun, November 26, 2000,
https://antarcticsun.usap.gov/pastIssues/2000-2001/2000_11_26.pdf (accessed 10/19/17).

[361] Antarctic Sun, November 26, 2000, https://antarcticsun.usap.gov/pastIssues/2000-2001/2000_11_26.pdf (accessed 10/19/17).

[362] Roger Highfield, "Antarctic Lake Isolated 40 Million Years To Be Explored", *The Electronic Telegraph* (9-21-1999). Copy available at: http://www.rense.com/general9/ant.htm (accessed 8/15/17).

[363] See Richard Hoagland and Mike Bara, "What is Happening at the South Pole?" http://www.enterprisemission.com/antarctica.htm (accessed 8/15/17).

[364] Kristan Hutchinson Sabbatini, "Soaring below Vostok," *The Antarctic Sun* (Feb 4, 2001), https://antarcticsun.usap.gov/pastIssues/2000-2001/2001_02_04.pdf (accessed 8/16/17).

[365] Dr Michael Studlinger is currently working for NASA at the Goodard Flight Center, in Maryland. His biography is available here: https://science.gsfc.nasa.gov/sed/bio/michael.studinger

[366] Kristan Hutchinson Sabbatini, "Soaring below Vostok," *The Antarctic Sun* (Feb 4, 2001), https://antarcticsun.usap.gov/pastIssues/2000-2001/2001_02_04.pdf (accessed 8/16/17).

[367] The Antarctic Sun (November 18, 2001) https://antarcticsun.usap.gov/pastIssues/2001-2002/2001_11_18.pdf (accessed 8/16/17).

[368] Richard Hoagland and Mike Bara, "What is Happening at the South Pole?" http://www.enterprisemission.com/antarctica.htm (accessed 8/15/17).

[369] Richard Hoagland and Mike Bara, "What is Happening at the South Pole?" http://www.enterprisemission.com/antarctica.htm (accessed 8/15/17).

[370] Henry Stevens, *Hitler's Suppressed and Still-Secret Weapons, Science and Technology* (Adventures Unlimited Press, 2007)p. *223.*

[371] Richard Hoagland and Mike Bara, "What is Happening at the South Pole?" http://www.enterprisemission.com/antarctica.htm (accessed 8/15/17).

[372] The Sun, "Shock claims massive ancient civilisation lies frozen beneath mile of Antarctic ice – and could even be Atlantis," https://www.thesun.co.uk/news/2380220/shock-claims-massive-civilisation-lies-frozen-beneath-a-mile-of-ice-in-the-south-pole/

[373] "Missing scientists mystery deepens in frozen Antarctica," http://www.foxnews.com/tech/2012/02/03/missing-scientists-mystery-deepens-in-frozen-antarctica.html (accessed 10/19/17).

[374] "Success! Russian Team Breaches Buried Antarctic Lake", https://www.livescience.com/18369-success-russian-team-breaches-buried-antarctic-lake-vostok.html (accessed 10/19/17).

[375] Michael Salla, "Military Abduction & Extraterrestrial Contact Treaty – Corey Goode Briefing Pt 2", http://exopolitics.org/military-abduction-extraterrestrial-contact-treaty-corey-goode-briefing-pt-2/ (accessed 10/21/2017).

[376] Corey Goode, "Endgame Part II: The Antarctic Atlantis & Ancient Alien Ruins", https://spherebeingalliance.com/blog/endgame-part-ii-the-antarctic-atlantis-and-ancient-alien-ruins.html (accessed 11/10/17).

[377] Corey Goode, "Endgame Part II: The Antarctic Atlantis & Ancient Alien Ruins", https://spherebeingalliance.com/blog/endgame-part-ii-the-antarctic-atlantis-and-ancient-alien-ruins.html (accessed 11/10/17).

[378] Charles Hapgood, *Earth's Shifting Crust: A Key To Some Basic Problems Of Earth Science* (Pantheon Books, 1958). Available online at: https://archive.org/stream/eathsshiftingcru033562mbp/eathsshiftingcru033562mbp_djvu.txt (accessed 11/10/17).

[379] Charles Hapgood, *Earth's Shifting Crust: A Key To Some Basic Problems Of Earth Science* (Pantheon Books, 1958). Available online at: https://archive.org/stream/eathsshiftingcru033562mbp/eathsshiftingcru033562mbp_djvu.txt (accessed 11/10/17).

[380] For discussion of the Oronteus Finaeus map, go to http://www.ancientdestructions.com/oronteus-finaeus-map-antarctica-fineus/ (accessed 11/10/17).

[381] Corey Goode, "Endgame Part II: The Antarctic Atlantis & Ancient Alien Ruins", https://spherebeingalliance.com/blog/endgame-part-ii-the-antarctic-atlantis-and-ancient-alien-ruins.html (accessed 11/10/17).

[382] See Michael Salla, "Impending Announcement of Ruins from Futuristic Civilization Found in Antarctica", http://exopolitics.org/impending-announcement-of-ruins-from-futuristic-civilization-found-in-antarctica/ (accessed 11/10/17).

[383] Corey Goode, "Endgame Part II: The Antarctic Atlantis & Ancient Alien Ruins", https://spherebeingalliance.com/blog/endgame-part-ii-the-antarctic-atlantis-and-ancient-alien-ruins.html (accessed 11/10/17).

[384] Corey Goode, "Endgame Part II: The Antarctic Atlantis & Ancient Alien Ruins", https://spherebeingalliance.com/blog/endgame-part-ii-the-antarctic-atlantis-and-ancient-alien-ruins.html (accessed 11/10/17).

[385] Arjun Walia, "DNA Analysis of Paracas Elongated Skulls Released: Unknown To Any Human, Primate, or Animal", http://www.collective-evolution.com/2014/02/12/dna-analysis-of-paracas-elongated-skulls-released-unknown-to-any-human-primate-or-animal/ (accessed 11/10/17).

[386] Corey Goode, "Endgame Part II: The Antarctic Atlantis & Ancient Alien Ruins", https://spherebeingalliance.com/blog/endgame-part-ii-the-antarctic-atlantis-and-ancient-alien-ruins.html (accessed 11/10/17).

[387] Corey Goode, "Endgame Part II: The Antarctic Atlantis & Ancient Alien Ruins", https://spherebeingalliance.com/blog/endgame-part-ii-the-antarctic-atlantis-and-ancient-alien-ruins.html (accessed 11/10/17).

CHAPTER TWELVE

[388] Kathryn Leishman forwarded a list of questions I had prepared to Congressman Nicholas Lampson and Dr. Rita Coleman, and she forwarded their responses on September 15, 2017.

[389] Email correspondence initiated by Kathryn Fleishman on my behalf with Brian.

[390] Corey Goode, "Latest Intel and Update", https://spherebeingalliance.com/blog/latest-intel-and-update.html (accessed 10/21/2017).

[391] See Michael Salla, "Secret Space Programs Battle over Antarctic Skies During Global Elite Exodus", http://exopolitics.org/secret-space-programs-battle-over-antarctic-skies-during-global-elite-exodus/ (accessed 11/10/2017).

[392] "Cosmic Disclosure: From Venus to Antarctica", Season 5, Episode 8, https://spherebeingalliance.com/blog/transcript-cosmic-disclosure-from-venus-to-antarctica.html (accessed 10/21/2017).

[393] "Cosmic Disclosure: From Venus to Antarctica", Season 5, Episode 8, https://spherebeingalliance.com/blog/transcript-cosmic-disclosure-from-venus-to-antarctica.html (accessed 10/21/2017).

[394] Chris Pash, "Scientists are closing in on warm caves under Antarctica which could support secret life", https://www.businessinsider.com.au/scientists-are-closing-in-on-warm-caves-under-antarctica-which-could-support-secret-life-2017-9

[395] "Cosmic Disclosure: From Venus to Antarctica", Season 5, Episode 8, https://spherebeingalliance.com/blog/transcript-cosmic-disclosure-from-venus-to-antarctica.html (accessed 10/21/2017).

[396] Robert McKie, "Scientists discover 91 volcanoes below Antarctic ice sheet", https://www.theguardian.com/world/2017/aug/12/scientists-discover-91-volcanos-antarctica

[397] "Cosmic Disclosure: From Venus to Antarctica", Season 5, Episode 8, https://spherebeingalliance.com/blog/transcript-cosmic-disclosure-from-venus-to-antarctica.html (accessed 10/21/2017).

[398] "Cosmic Disclosure: From Venus to Antarctica", Season 5, Episode 8, https://spherebeingalliance.com/blog/transcript-cosmic-disclosure-from-venus-to-antarctica.html (accessed 10/21/2017).

[399] Corey Goode, "Latest Intel and Update", https://spherebeingalliance.com/blog/latest-intel-and-update.html (accessed 10/21/2017).

[400] "Cosmic Disclosure: Deeper Disclosures from William Tompkins," https://spherebeingalliance.com/blog/transcript-cosmic-disclosure-deeper-disclosures-from-william-tompkins.html (accessed 11/10/2017).

[401] "Cosmic Disclosure: From Venus to Antarctica", Season 5, Episode 8, https://spherebeingalliance.com/blog/transcript-cosmic-disclosure-from-venus-to-antarctica.html (accessed 10/21/2017).

[402] "Cosmic Disclosure: From Venus to Antarctica", Season 5, Episode 8, https://spherebeingalliance.com/blog/transcript-cosmic-disclosure-from-venus-to-antarctica.html (accessed 10/21/2017).

[403] "Cosmic Disclosure: From Venus to Antarctica", Season 5, Episode 8, https://spherebeingalliance.com/blog/transcript-cosmic-disclosure-from-venus-to-antarctica.html (accessed 10/21/2017).

[404] "Cosmic Disclosure: From Venus to Antarctica", Season 5, Episode 8, https://spherebeingalliance.com/blog/transcript-cosmic-disclosure-from-venus-to-antarctica.html (accessed 10/21/2017).

[405] William Tompkins interviewed on *Cosmic Disclosure*, February 21 2017. https://spherebeingalliance.com/blog/transcript-cosmic-disclosure-deeper-disclosures-from-william-tompkins.html (accessed 11/9/17).

[406] Michael Salla, "Secret NRO Space Stations to be Revealed in Limited Disclosure Plan", http://exopolitics.org/secret-nro-space-stations-to-be-revealed-in-limited-disclosure-plan/ (accessed 10/21/2017).

[407] Corey Goode, "Latest Intel and Update", https://spherebeingalliance.com/blog/latest-intel-and-update.html (accessed 10/21/2017).

[408] See Michael Salla, "Secret Space Programs Battle over Antarctic Skies During Global Elite Exodus", http://exopolitics.org/secret-space-programs-battle-over-antarctic-skies-during-global-elite-exodus/ (accessed 11/10/2017).

[409] Coast to Coast Radio interview, March 21, 2016, https://www.coasttocoastam.com/show/2016/03/21 (accessed 10/21/2017).

[410] Michael Salla, "Secret NRO Space Stations to be Revealed in Limited Disclosure Plan", http://exopolitics.org/secret-nro-space-stations-to-be-revealed-in-limited-disclosure-plan/ (accessed 10/21/2017).

[411] Michael Salla, "Secret Space Program Conferences discuss full disclosure & humanity's future", http://exopolitics.org/secret-space-program-conferences-discuss-full-disclosure-humanitys-future/ (accessed 10/21/2017).

[412] The Declaration of San Carlos de Bariloche: Joint Declaration by the President and President Frondizi of Argentina. http://www.presidency.ucsb.edu/ws/?pid=12127 (accessed 10/21/2017).

[413] See Harry Cooper, Hitler in Argentina: The Documented Truth of Hitler's Escape from Berlin (Createspace 2014).

[414] Corey Goode, "Latest Intel and Update", https://spherebeingalliance.com/blog/latest-intel-and-update.html (accessed 10/21/2017).

[415] Michael Salla, "Alliance of Secret Space Programs Adopts Scaled Back Alien-UFO Disclosure Plan," http://exopolitics.org/alliance-of-secret-space-programs-adopts-scaled-back-alien-ufo-disclosure-plan/ (accessed 10/21/2017).

[416] Corey Goode, "Latest Intel and Update", https://spherebeingalliance.com/blog/latest-intel-and-update.html (accessed 10/21/2017).

[417] Corey Goode, "Latest Intel and Update", https://spherebeingalliance.com/blog/latest-intel-and-update.html (accessed 10/21/2017).

[418] Robert E. McElwaine, "Russian Cosmosphere: Operational Star Wars Defense System", http://www.bibliotecapleyades.net/sociopolitica/esp_sociopol_firesky_01.htm (accessed 10/21/2017).

[419] See "Cosmic Disclosure: Antarctica: The Process for Disclosure", https://spherebeingalliance.com/blog/transcript-cosmic-disclosure-antarctica-the-process-for-disclosure.html (accessed 11/11/2017).

[420] *Cosmic Disclosure*, "Antarctica: The Process for Disclosure", Season 7, Episode 8 https://spherebeingalliance.com/blog/transcript-cosmic-disclosure-antarctica-the-process-for-disclosure.html (accessed 10/22/17).

[421] "Cosmic Disclosure: Antarctica: The New Area 51", https://spherebeingalliance.com/blog/transcript-cosmic-disclosure-antarctica-the-new-area-51.html (accessed 11/11/2017).

[422] "Cosmic Disclosure: Antarctica: The New Area 51", https://spherebeingalliance.com/blog/transcript-cosmic-disclosure-antarctica-the-new-area-51.html (accessed 11/11/2017).

[423] Interviews available online at: http://projectcamelot.org/pete_peterson.html (accessed 11/11/2017).

[424] "Cosmic Disclosure: UFOs under Antarctica and the Five-Fingered Mystery", https://spherebeingalliance.com/blog/transcript-cosmic-disclosure-ufos-under-antarctica-and-the-five-fingered-mystery.html (accessed 11/11/2017).

[425] *Cosmic Disclosure*, "Antarctica: The Process for Disclosure", Season 7, Episode 8 https://spherebeingalliance.com/blog/transcript-cosmic-disclosure-antarctica-the-process-for-disclosure.html (accessed 10/22/17).

[426] Michael Salla, "Visit to Antarctica Confirms Discovery of Flash Frozen Alien Civilization", http://exopolitics.org/visit-to-antarctica-confirms-discovery-of-flash-frozen-alien-civilization/ (accessed 10/21/2017).

[427] Michael Salla, "Sitchin's Sumerian Text Translations Contrived by Illuminati to Promote False Alien Religion", http://exopolitics.org/sitchins-sumerian-text-translations-contrived-by-illuminati-to-promote-false-alien-religion/ (accessed 10/21/2017).

[428] Corey Goode, "Endgame Part II: The Antarctic Atlantis & Ancient Alien Ruins", https://spherebeingalliance.com/blog/endgame-part-ii-the-antarctic-atlantis-and-ancient-alien-ruins.html (accessed 10/21/2017)

[429] Charles Hapgood, Path of the Pole (Adventures Unlimited Press, 1999 [1970]).

[430] The Guardian, "Scientists discover 91 volcanoes below Antarctic ice sheet", https://www.theguardian.com/world/2017/aug/12/scientists-discover-91-volcanos-antarctica#img-1 (accessed 1/31/2018).

[431] NASA, "Temperature Trends in Antarctica", https://svs.gsfc.nasa.gov/3575 (accessed 1/31/2018).

[432] Cosmic Disclosure with David Wilcock, Season 9, Episode 9 (January 23, 2018) https://spherebeingalliance.com/blog/transcript-cosmic-disclosure-hybrid-creatures-and-secret-bases.html

[433] National Geographic, "What the World Would Look Like if All the Ice Melted" https://www.nationalgeographic.com/magazine/2013/09/rising-seas-ice-melt-new-shoreline-maps/(accessed 1/31/2018).

[434] Albert Einstein Foreword in Charles Hapgood, *The Earth's Shifting Crust* (Pantheon Books, 1958) p. 1. Available online at: https://archive.org/stream/eathsshiftingcru033562mbp/eathsshiftingcru033562mbp_djvu.txt (accessed 1/31/18).

[435] Chris Pash, "Scientists are closing in on warm caves under Antarctica which could support secret life", https://www.businessinsider.com.au/scientists-are-closing-in-on-warm-caves-under-antarctica-which-could-support-secret-life-2017-9

CHAPTER THIRTEEN

[436] "Cosmic Disclosure: Antarctica: The Process for Disclosure," Season 7, Episode 8 https://spherebeingalliance.com/blog/transcript-cosmic-disclosure-antarctica-the-process-for-disclosure.html (accessed 10/21/17)

[437] *Voice of America*, "Mars Once Had Oxygen-Rich Atmosphere", https://www.voanews.com/a/mars-oxygen/1713223.html (accessed 10/21/17)

[438] "Super Earth", *Cosmic Disclosure*, 28 May, 2016, Season 4, Episode 1 https://spherebeingalliance.com/blog/transcript-cosmic-disclosure-super-earth.html (accessed 10/21/17)

[439] Thomas Van Flandern, "The Exploded Planet Hypothesis 2000" http://tinyurl.com/y9sveesj (accessed 10/21/17)

[440] "Mars Exploration," https://www.cia.gov/library/readingroom/docs/CIA-RDP96-00788R001900760001-9.pdf

[441] See Martin Gardner, "The Great Stone Face and Other Nonmysteries", http://archive.li/UmiAV#selection-227.0-231.14 (accessed 10/21/17)

[442] John Connolly, "The Secret History of the National Enquirer", http://dujour.com/news/national-enquirer-history-scandal/ (accessed 10/21/17)

[443] Richard C. Hoagland, *The Monuments of Mars: A City on the Edge of Forever* (Frog Books; 5th ed. edition, 2001)

[444] Ingo Swann, *Penetration: The Question of Extraterrestrial and Human Telepathy* (Ingo Swann Books; 1998).

[445] Ingo Swann, *Penetration,* chapter 5.

[446] "Mars Exploration," https://www.cia.gov/library/readingroom/docs/CIA-RDP96-00788R001900760001-9.pdf (accessed 10/21/17)

[447] "Mars Exploration," https://www.cia.gov/library/readingroom/docs/CIA-RDP96-00788R001900760001-9.pdf (accessed 10/21/17)

[448] "Mars Exploration," https://www.cia.gov/library/readingroom/docs/CIA-RDP96-00788R001900760001-9.pdf (accessed 10/21/17)

[449] "Mars Exploration,"

https://www.cia.gov/library/readingroom/docs/CIA-RDP96-00788R001900760001-9.pdf (accessed 10/21/17)

[450] "Mars Exploration," https://www.cia.gov/library/readingroom/docs/CIA-RDP96-00788R001900760001-9.pdf (accessed 10/21/17)

[451] "Mars Exploration," https://www.cia.gov/library/readingroom/docs/CIA-RDP96-00788R001900760001-9.pdf (accessed 10/21/17)

[452] "Mars Exploration," https://www.cia.gov/library/readingroom/docs/CIA-RDP96-00788R001900760001-9.pdf (accessed 10/21/17)

[453] An extract of the lecture was published on Youtube, "Joe McMoneagle - Remote viewing of Mars (2004)", https://youtu.be/HlLq7KDU2HY (accessed 11/07/17)

[454] Holmes Skip Atwater, "Mars", http://www.skipatwater.com/training.html#mars (accessed 11/12/17)

[455] Holmes Skip Atwater, "Remote Viewing Mars," https://youtu.be/t8UG0Asa7jY (accessed 11/12/17)

[456] Chris Pash, "Scientists are closing in on warm caves under Antarctica which could support secret life", https://www.businessinsider.com.au/scientists-are-closing-in-on-warm-caves-under-antarctica-which-could-support-secret-life-2017-9 (accessed 10/22/17)

[457] Charles Hapgood, *Path of the Pole* (Adventures Unlimited Press, 2015 [1970])

[458] Available online at: https://wireofinformation.wordpress.com/tag/einstein-foreword-to-earths-shifting-crust/ (accessed 10/23/17)

[459] Available online at: https://wireofinformation.wordpress.com/tag/einstein-foreword-to-earths-shifting-crust/ (accessed 10/23/17)

[460] "Cosmic Disclosure: Testimony on Pyramids and Underground Cities", https://spherebeingalliance.com/blog/transcript-cosmic-disclosure-testimony-on-pyramids-and-underground-cities.html (accessed 11/11/17)

[461] See Michael Salla, "Moon is Artificial & Arrived with Refugees from Destroyed Planet in Asteroid Belt", http://exopolitics.org/moon-is-artificial-arrived-with-refugees-from-destroyed-planet-in-asteroid-belt/ (accessed 10/22/17)

[462] Laura Geggel, "City-Size Lunar Lava Tube Could House Future Astronaut Residents", https://www.livescience.com/60733-moon-lava-tube-could-shelter-astronauts.html (accessed 10/22/17)

[463] Immanuel Velikovsky, "The Earth Without the Moon", http://www.varchive.org/itb/sansmoon.htm (accessed 10/22/17)

[464] *Cosmic Disclosure*, "Antarctica: The Process for Disclosure", Season 7, Episode 8 https://spherebeingalliance.com/blog/transcript-cosmic-disclosure-antarctica-the-process-for-disclosure.html (accessed 10/22/17)

[465] *Cosmic Disclosure*, "Antarctica: The Process for Disclosure", Season 7, Episode 8 https://spherebeingalliance.com/blog/transcript-cosmic-disclosure-antarctica-the-process-for-disclosure.html (accessed 10/22/17)

[466] *Cosmic Disclosure*, "Antarctica: The Process for Disclosure", Season 7, Episode 8 https://spherebeingalliance.com/blog/transcript-cosmic-disclosure-antarctica-the-process-for-disclosure.html (accessed 10/22/17)

[467] *Cosmic Disclosure*, "Antarctica: The Process for Disclosure", Season 7, Episode 8 https://spherebeingalliance.com/blog/transcript-cosmic-disclosure-antarctica-the-process-for-disclosure.html (accessed 10/22/17)

[468] "Mars Exploration," https://www.cia.gov/library/readingroom/docs/CIA-RDP96-00788R001900760001-9.pdf (accessed 10/21/17)

CHAPTER FOURTEEN

[469] *Cosmic Disclosure*, "Antarctica: The Process for Disclosure", Season 7, Episode 8 https://spherebeingalliance.com/blog/transcript-cosmic-disclosure-antarctica-the-process-for-disclosure.html (accessed 10/22/17)

[470] Manetho, Book I, p. 2, https://archive.org/stream/manethowithengli00maneuoft/manethowithengli00maneuoft_djvu.txt (accessed 10/22/17)

[471] Gary Lynch and Richard Granger, *Big Brain: The Origins and Future of Human Intelligence* (St. Martin's Press, 2008)

[472] Gary Lynch & Richard Granger, "What Happened to the Hominids Who May Have Been Smarter Than Us?", Discover Magazine, http://discovermagazine.com/2009/the-brain-2/28-what-happened-to-hominids-who-were-smarter-than-us (accessed 10/23/17)

[473] For description of the Great Flood in Sumerian texts, see "The Great Flood: Sumerian version", http://www.livius.org/articles/misc/great-flood/flood2/

[474] "The Sumerian king list: translation", http://etcsl.orinst.ox.ac.uk/section2/tr211.htm (accessed 10/22/17)

[475] "The Sumerian king list: translation", http://etcsl.orinst.ox.ac.uk/section2/tr211.htm (accessed 10/22/17)

[476] "The Sumerian king list: translation", http://etcsl.orinst.ox.ac.uk/section2/tr211.htm (accessed 10/22/17)

[477] For biblical passages referring to the age of these pre-deluvial figures, see "Longevity", https://bible.knowing-jesus.com/topics/Longevity (accessed 11/11/17)

[478] For biblical passages referring to the age of these pre-deluvial figures, see "Longevity", https://bible.knowing-jesus.com/topics/Longevity (accessed 11/11/17)

[479] Graham Hancock, *Magicians of the Gods* (A Thomas Dunne Book for St. Martin's Griffin, 2017)

[480] *Cosmic Disclosure*, "Antarctica: The Process for Disclosure", Season 7, Episode 8 https://spherebeingalliance.com/blog/transcript-cosmic-disclosure-antarctica-the-process-for-disclosure.html (accessed 10/22/17)

[481] *Cosmic Disclosure*, "Antarctica: The Process for Disclosure", Season 7, Episode 8 https://spherebeingalliance.com/blog/transcript-cosmic-disclosure-antarctica-the-process-for-disclosure.html (accessed 10/22/17)

[482] *Cosmic Disclosure*, "Antarctica: The Process for Disclosure", Season 7, Episode 8 https://spherebeingalliance.com/blog/transcript-cosmic-disclosure-antarctica-the-process-for-disclosure.html (accessed 10/22/17)

[483] *Cosmic Disclosure*, "Antarctica: The Process for Disclosure", Season 7, Episode 8 https://spherebeingalliance.com/blog/transcript-cosmic-disclosure-antarctica-the-process-for-disclosure.html (accessed 10/22/17)

[484] Gary Lynch & Richard Granger, "What Happened to the Hominids Who May Have Been Smarter Than Us?", Discover Magazine, http://discovermagazine.com/2009/the-brain-2/28-what-happened-to-hominids-who-were-smarter-than-us (accessed 10/23/17)

[485] *Cosmic Disclosure*, "Antarctica: The Process for Disclosure", Season 7, Episode 8 https://spherebeingalliance.com/blog/transcript-cosmic-disclosure-antarctica-the-process-for-disclosure.html (accessed 10/22/17)

[486] "The Fallen Angels Imprisoned in Antarctica and are still Alive!" https://www.youtube.com/watch?v=Bez4DKZI7yU&feature=youtu.be (accessed 11/1/17)

[487] "The Book of Enoch," http://www.markfoster.net/rn/texts/AllBooksOfEnoch.pdf (accessed 11/1/17)

[488] "The Book of Enoch," http://www.markfoster.net/rn/texts/AllBooksOfEnoch.pdf (accessed 11/1/17)

[489] Available online at: https://archive.org/stream/eathsshiftingcru033562mbp/eathsshiftingcru033562mbp_djvu.txt (accessed 11/1/17)

[490] "The Book of Enoch," http://www.markfoster.net/rn/texts/AllBooksOfEnoch.pdf (accessed 11/1/17)

[491] "Whistleblower reveals multiple secret space programs concerned about new alien visitors", http://exopolitics.org/whistleblower-reveals-multiple-secret-space-programs-concerned-about-new-alien-visitors/ (accessed 11/1/17)

[492] "Visit to Antarctica Confirms Discovery of Flash Frozen Alien Civilization", http://exopolitics.org/visit-to-antarctica-confirms-discovery-of-flash-frozen-alien-civilization/

[493] *Cosmic Disclosure*, "Antarctica: The Process for Disclosure", Season 7, Episode 8 https://spherebeingalliance.com/blog/transcript-cosmic-disclosure-antarctica-the-process-for-disclosure.html (accessed 10/22/17)

[494] https://www.facebook.com/BlueAvians/posts/1453552121609246 (accessed 11/1/17)

[495] Alex Collier, "An Andromedan Perspective on Galactic History", http://www.exopoliticsjournal.com/vol-2/vol-2-2-Collier.pdf (accessed 11/1/17)

[496] *Cosmic Disclosure*, "Antarctica: The Process for Disclosure", Season 7, Episode 8 https://spherebeingalliance.com/blog/transcript-cosmic-disclosure-antarctica-the-process-for-disclosure.html (accessed 10/22/17)

[497] Corey Goode, "Endgame Part II: The Antarctic Atlantis & Ancient Alien Ruins" https://spherebeingalliance.com/blog/endgame-part-ii-the-antarctic-atlantis-and-ancient-alien-ruins.html (accessed 11/1/17)

[498] *Cosmic Disclosure*, "Antarctica: The Process for Disclosure", Season 7, Episode 8 https://spherebeingalliance.com/blog/transcript-cosmic-disclosure-antarctica-the-process-for-disclosure.html (accessed 10/22/17)

CHAPTER FIFTEEN

[499] "National Industrial Security Program Operating Manual:" DOD 5220.22-M-Sup. 1, February 1995. 1-1-2: https://www.fas.org/sgp/library/nispom_sup.pdf (accessed December 2013)

[500] Ibid. 3-1-2 & A-4, https://www.fas.org/sgp/library/nispom_sup.pdf (accessed December 2013).

[501] "Report of the Commission on Protecting and Reducing Government Secrecy," (Senate Document. 105-2 – 3 December 1997), 26, http://www.gpo.gov/fdsys/pkg/GPO-CDOC-105sdoc2/pdf/GPO-CDOC-105sdoc2-7.pdf (accessed December 2013)

[502] "Special Access Program Supplement to the National Industrial Security," (Draft 29 May 1992). 3-1-5, *www.fas.org/sgp/library/nispom/sapsup-draft92.pdf* (accessed December 2013)

[503] "Report of the Commission on Protecting and Reducing Government Secrecy," https://www.fas.org/sgp/library/moynihan/chap2.pdf (accessed December 2013)

[504] Tim Cook, *Blank Check: The Pentagon's Black Budget* (Grand Central Publishing, 1990).

[505] For discussion of the deep black budget, see Michael Salla, "The Black Budget Report: An Investigation into the CIA's 'Black Budget' and the Second Manhattan Project," http://exopolitics.org/Report-Black-Budget.htm

[506] "Department of Defense (DoD) Releases Fiscal Year 2018 Budget Proposal," http://www.defense.gov/News/News-Releases/News-Release-View/Article/1190216/dod-releases-fiscal-year-2018- budget-proposal (accessed 12/20/17)

[507] Stars and Stripes, "Report: 44,000 'unknown' military personnel stationed around the world" https://www.stripes.com/report-44-000-unknown-military-personnel-stationed-around-the-world-1.501292 (accessed 12/20/17)

[508] "Company Overview of ITT Antarctic Services, Inc." https://www.bloomberg.com/research/stocks/private/snapshot.asp?privcapId=34119998 (accessed 11/01/17)

[509] "Cosmic Disclosure: Antarctica: The Process for Disclosure", https://spherebeingalliance.com/blog/transcript-cosmic-disclosure-antarctica-the-process-for-disclosure.html (accessed 11/11/17)

[510] See "Cosmic Disclosure: UFOs under Antarctica and the Five-Fingered Mystery", https://spherebeingalliance.com/blog/transcript-cosmic-disclosure-ufos-under-antarctica-and-the-five-fingered-mystery.html

[511] Private interview with William Tompkins, July 30, 2017.

[512] Jonathan Shikes, "Life in Antarctica is cold — but bloggers there can still get burned" http://www.westword.com/news/life-in-antarctica-is-cold-but-bloggers-there-can-still-get-burned-5105529 (accessed 11/01/17)

[513] "Pole postcards...old and new", http://www.southpolestation.com/postcard/index.html (accessed 11/01/17)

[514] AECOM http://www.aecom.com/markets/government/energy/ (accessed 11/05/17)

[515] "Pole postcards...old and new", http://www.southpolestation.com/postcard/index.html (accessed 11/01/17)

[516] http://www.referenceforbusiness.com/history2/12/AECOM-Technology-Corporation.html (accessed 11/01/17)

[517] "Company Overview of ITT Antarctic Services, Inc." https://www.bloomberg.com/research/stocks/private/snapshot.asp?privcapId=34119998 (accessed 11/01/17)

[518] Rodney E. Gray, Support operations of ITT/Antarctic Services, Inc.http://tinyurl.com/yb44fh6e (accessed 11/12/17)

[519] Jonathan Shikes, "Life in Antarctica is cold — but bloggers there can still get burned" http://www.westword.com/news/life-in-antarctica-is-cold-but-bloggers-there-can-still-get-burned-5105529 (accessed 11/01/17)

[520] "Janet Airline / EG&G / JT3", http://www.dreamlandresort.com/info/janet.html (accessed 11/02/17)

[521] Gene Huff, The Lazar Synopsis, http://www.otherhand.org/home-page/area-51-and-other-strange-places/bluefire-main/bluefire/the-bob-lazar-corner/the-lazar-synopsis/ (accessed 11/11/17)

[522] "Special Access Program Supplement to the National Industrial Security," (Draft 29 May 1992). 3-1-5, *https://fas.org/sgp/library/nispom/sapsup-draft92.pdf* (11/01/17)

[523] https://web.archive.org/web/20120208073024/http://rpsc.raytheon.com/ (accessed 11/01/17)

[524] Raytheon, Wikipedia, https://en.wikipedia.org/wiki/Raytheon (accessed 11/01/17)

[525] Raytheon, "Who We Are", https://www.raytheon.com/ourcompany/ (accessed 11/02/17)

[526] William Maarkin, "We are SAPs: forty companies currently working on 'special access programs'" https://williamaarkin.wordpress.com/2012/05/31/we-are-saps-forty-companies-currently-working-on-special-access-programs/ (accessed 11/01/17)

[527] Richard Hoagland and Mike Bara, "What is Happening at the South Pole?" http://www.enterprisemission.com/antarctica.htm (accessed 8/15/17)

[528] National Science Foundation, "Auditors Report", https://www.scribd.com/document/1001101/National-Science-Foundation-06-1-004-RPSC (accessed 11/01/17)

[529] "Raytheon Awarded a One-Year Extension to United States Antarctic Program Support Contract", http://raytheon.mediaroom.com/index.php?item=1529 (accessed 11/01/17)

[530] "Raytheon censors Antarctic bloggers", https://dearkitty1.wordpress.com/2009/10/08/raytheon-censors-antarctic-bloggers-2/ (accessed 11/02/17)

[531] Jeffrey Mervis, "Updated: NSF Picks Lockheed for Huge Antarctic Support Contract", http://www.sciencemag.org/news/2011/12/updated-nsf-picks-lockheed-huge-antarctic-support-contract (accessed 11/01/17)

[532] Wikipedia, Lockheed Martin, https://en.wikipedia.org/wiki/Lockheed_Martin (accessed 11/01/17)

[533] Clarence L. "Kelly" Johnson, *Kelly: More Than My Share of It* (All Smithsonian Books, 1989)

[534] William Tompkins, *Selected by Extraterrestrials* (Createspace, 2015) 427.

[535] Private Interview with William Tompkins, April 17, 2017. To be published in the second instalment of William Tompkins autobiography, (forthcoming 2018).

[536] Private Interview with William Tompkins, April 17, 2017 To be published in the second instalment of William Tompkins autobiography, (forthcoming 2018).

[537] For description of USAPs at Lockheed, see Ben Rich, *Skunk Works: A Personal Memoir of My Years of Lockheed* (Little, Brown and Company, 2013) 96-98

[538] Jeffrey Richelson, "The Secret History of the U-2 - and Area 51," http://nsarchive.gwu.edu/NSAEBB/NSAEBB434/ (accessed 11/2/17).

[539] Ben Rich, *Skunk Works: A Personal Memoir of My Years at Lockheed* (Back Bay Books, 1996)

[540] National Science Foundation, "New manager for US Antarctic Program logistics contract" https://www.nsf.gov/news/news_summ.jsp?cntn_id=189574&org=NSF&from=news (accessed 11/2/17).

[541] Leidos, "A Visionary Leader with a Lasting Legacy", https://www.leidos.com/about/history/beyster (accessed 11/2/17).

[542] Amrita Jayakumar, "One year later: The tale of SAIC and Leidos", https://www.washingtonpost.com/business/capitalbusiness/one-year-later-saic-and-leidos/2014/09/26/d1fefd68-4273-11e4-b437-1a7368204804_story.html (accessed 11/2/17).

[543] Nick Wakeman, "What's Leidos getting for $5B?" https://fcw.com/articles/2016/01/28/wakeman-leidos-analysis.aspx (accessed on 11/2/17).

[544] Richard Boylan, "Inside Revelations on the UFO Cover-Up", http://ufoevidence.org/documents/doc1861.htm (accessed 11/2/17).

[545] See William Tompkins, *Selected by Extraterrestrials,* pp. 12, 224, 312-13

[546] William Tompkins, *Selected by Extraterrestrials,* pp. 312-13.

[547] William Tompkins, *Selected by Extraterrestrials,* pp. 12.

[548] Private Interview with Admiral Bobby Ray Inman, December 1, 2016.

[549] For a full list of companies that Admiral Bobby Ray Inman has served on go to: http://www.nndb.com/people/392/000058218/(accessed 11/3/17).

[550] For his retirement notice see https://web.archive.org/web/20040209210833/http://www.saic.com:80/news/20 03/oct/09.html (accessed 11/12/17).

[551] James Bamford, *The Shadow Factory, The NSA from 9/11 to the Eavesdropping on America* (Anchor 2009) p. 201.

[552] See Leuren Moret, "Nuclear Weapons Stealth Takeover" *Global Research* (9/9/2004) https://www.globalresearch.ca/articles/MOR409A.html (accessed 11/2/17).

[553] Kristan Hutchinson, "Antarctica: Almost Out of this World"", *The Antarctic Sun,* https://antarcticsun.usap.gov/pastIssues/2002-2003/2002_12_29.pdf (accessed 11/11/17).

[554] The SAIC meteorite study was formally part of the NASA Johnson Space Center's exploration offices. See Kristan Hutchinson, "Antarctica: Almost Out of this World", *The Antarctic Sun,* https://antarcticsun.usap.gov/pastIssues/2002-2003/2002_12_29.pdf (accessed 11/11/17)

[555] An extract of the lecture was published on Youtube, "Joe McMoneagle - Remote viewing of Mars (2004)", https://youtu.be/HlLq7KDU2HY (accessed 11/07/17)

[556] Federation of Atomic Scientists, "STAR GATE [Controlled Remote Viewing]" https://fas.org/irp/program/collect/stargate.htm (accessed 11/13/17)

[557] Holmes Skip Atwater, "Mars", http://www.skipatwater.com/training.html#mars (accessed 11/12/17)

[558] See Dr Richard Wiseman and and Dr Julie Milton, "Experiment One of the SAIC Remote Viewing Program: A critical re-evaluation" http://www.richardwiseman.com/resources/SAICcrit.pdf (accessed 11/12/17)

[559] Private Skype communication with Corey Goode, July 31, 2017.

[560] Private Skype communication with Corey Goode, July 31, 2017.

[561] Michael Salla and Corey Goode, "Illegal Military Research and Development in Antarctica", http://exopolitics.org/illegal-military-research-and-development-in-antarctica/

[562] Private Skype communication with Corey Goode, July 31, 2017.

[563] In October 2011, ITT Corporation's defense business was spun off as the new public traded company Exelis. In 2015, Exelis was purchased by the Harris Corporation for $4.75 billion.

[564] Private Interview with William Tompkins, April 17, 2017 To be published in the second installment of William Tompkins autobiography, (forthcoming 2018).

[565] Private Interview with William Tompkins, April 17, To be published in the second installment of William Tompkins autobiography, (forthcoming 2018). 2017

[566] Robin K Burrows, "A Contractor's Guide to the Freedom of Information Act", https://watttieder.com/resources/articles/a-contractor%27s-guide-to-the-freedom-of-information-act (accessed on 11/5/17)

[567] See Joel D. Hesch, "Whistleblower Protection and Rewards for Defense Contractors & Subcontractors" http://www.howtoreportfraud.com/blog/whistleblower-protection-and-rewards-for-defense-contractors-subcontractors/ (accessed on 11/5/17)

[568] An interview with Representative Sherwood Boehlert, the leader of the Congressional Delegation, was published in the Antarctic Sun, "Congressional delegation visits, praises program" https://antarcticsun.usap.gov/pastIssues/2002-2003/2003_01_26.pdf (accessed 11/7/17)

[569] Congressman Nicholas Lehman and Dr Rita Colwell, head of the National Science Foundation at the time, were asked a series of questions I had prepared. Kathryn Leishman interviewed them on my behalf in September 2017

[570] Leidos 2016 Annual Report can be downloaded from its website at: http://investors.leidos.com/phoenix.zhtml?c=193857&p=irol-sec (accessed 11/2/17)

[571] "Harris: 2017 Annual Report" https://www.harris.com/sites/default/files/2017-harris-annual-report.pdf (accessed on 11/5/17)

[572] For analysis of the black budget, see Michael Salla, "The Black Budget Report: An Investigation into the CIA's 'Black Budget' and the Second Manhattan Project" http://exopolitics.org/Report-Black-Budget.htm (accessed 11/5/17)

[573] See Michael Salla, "The Black Budget Report: An Investigation into the CIA's 'Black Budget' and the Second Manhattan Project", http://exopolitics.org/archived/Report-Black-Budget.htm (accessed 11/13/17)

[574] See NASA, "Antarctic Meteorites" https://curator.jsc.nasa.gov/antmet/index.cfm (accessed 11/6/17)

CHAPTER SIXTEEN

[575] For further discussion of the Interplanetary Corporate Conglomerate and Solar Warden see Michael Salla, *Insiders Reveal Secret Space Programs and Extraterrestrial Alliances* (Exopolitics Institute,. 2015)

[576] *Cosmic Disclosure*, "Antarctica: The Process for Disclosure", Season 7, Episode 8 https://spherebeingalliance.com/blog/transcript-cosmic-disclosure-antarctica-the-process-for-disclosure.html (accessed 10/22/17)

[577] *Cosmic Disclosure*, "Antarctica: The Process for Disclosure", Season 7, Episode 8 https://spherebeingalliance.com/blog/transcript-cosmic-disclosure-antarctica-the-process-for-disclosure.html (accessed 10/22/17)

[578] *Cosmic Disclosure*, "Antarctica: The Process for Disclosure", Season 7, Episode 8 https://spherebeingalliance.com/blog/transcript-cosmic-disclosure-antarctica-the-process-for-disclosure.html (accessed 10/22/17)

[579] William Tompkins, Volume 2: Selected by Extraterrestrials, forthcoming 2018

[580] See Albert Speer, *Infiltration*, pp. 301-302.

[581] For statistics on missing people from different countries go to: "Missing People Worldwide" http://www.mcatracing.co.uk/missing-people-worldwide.htm (accessed on 11/7/17)

[582] Wikipedia, "Restorative Justice," https://en.wikipedia.org/wiki/Restorative_justice (accessed on 11/7/17)

[583] Alex Collier, "The Rest of the Galactic Hierarchy... and the Rest of the Story" https://www.bibliotecapleyades.net/sumer_anunnaki/reptiles/reptiles33.htm (accessed 12/17/17)

[584] See Michael Salla, "Secret Mars Colonies Trade with up to 900 Extraterrestrial Civilizations", http://exopolitics.org/secret-mars-colonies-trade-with-up-to-900-extraterrestrial-civilizations/

INDEX

C

Made in the USA
San Bernardino, CA
28 April 2018